Stethoscope

Stethoscope

The Making of a Medical Icon

Anna Harris and Tom Rice

Reaktion Books

For Arthur and Bastian

Published by Reaktion Books Ltd
Unit 32, Waterside
44–48 Wharf Road
London N1 7UX, UK
www.reaktionbooks.co.uk

First published 2022
Copyright © Anna Harris and Tom Rice 2022
The author order is alphabetical representing equal co-authorship

Printed and bound in Great Britain by
TJ Books Ltd, Padstow, Cornwall

A catalogue record for this book is available from the British Library

ISBN 978 1 78914 633 2

Contents

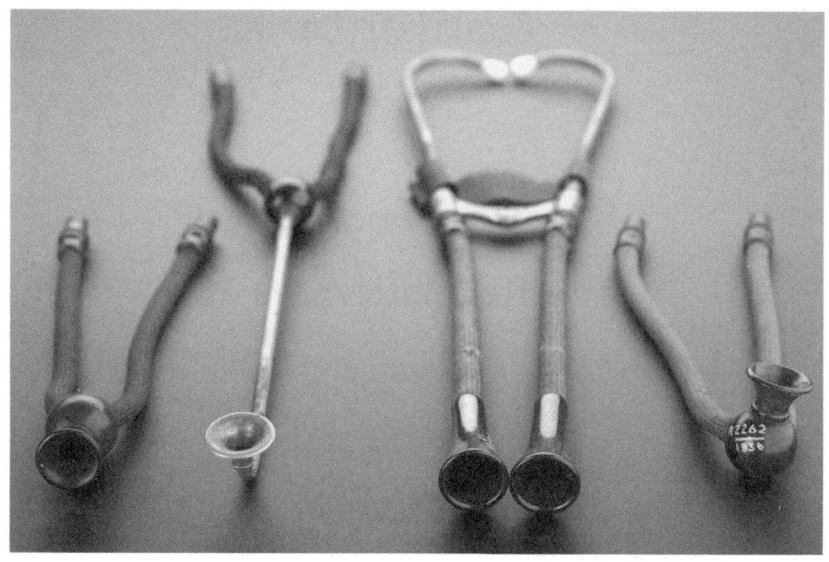

Early models of binaural stethoscope, involving the use of both ears, an elaboration upon monaural (one-eared) versions of the instrument.

Introduction

This book explores the past, present and future of an instrument close to our hearts. Every day the stethoscope is involved in thousands of interactions between healthcare professionals and patients all over the world. Most of us will be examined by a stethoscope at some point in our lives. It has been woven into medical routine and so is an object with which we have all had personal contact. Many children discover stethoscopes at a young age in toy doctors' kits. It also continually appears in the public domain, on screens, posters and leaflets, as a symbol of medicine itself. But how have stethoscopes become part of our lives? What makes the instrument so iconic of the medical profession? Why is it both such an ordinary and such a charismatic object?

Invented in the nineteenth century, the stethoscope was a product of the drive to establish rationalism and science as the basis of medicine. The monaural (one-eared) stethoscope represented a transformative diagnostic advance. It enabled doctors to use their ears to 'see' inside the body, especially the heart and lungs, of a living patient. They no longer had to rely on patients' stories alone. Initially resisted by parts of the medical establishment, by the end of the nineteenth century the ability to detect and interpret a variety of murmurs, crackles and other sounds had become a vital skill for the modern doctor in places where Western medicine was practised. Wielding a (by this time often binaural) stethoscope was a mark of expertise and a badge of professional membership.

Today, the conventional stethoscope might seem like a relic of a bygone era, outmoded by digital versions or rendered obsolete by portable ultrasound machines and artificial intelligence. Given the ever-increasing technological power and sophistication of the medical world, what, if anything, do stethoscopes continue to offer? Where and why do doctors persist in using them? Why do they still hold such public appeal as symbols of medicine? In this book we address these questions. To do so we draw on our experience as anthropologists who, as we explain in more detail below, have both carried out studies of stethoscopic listening. We draw on insights gained from our field-work, as well as on library-based research, combining our knowledge to provide a cultural history of the stethoscope that encompasses use of the stethoscope in contemporary times and a speculative future.

Why is it important to revisit the stethoscope now? We believe that the instrument is at the heart of questions about what kind of medicine we want in the contemporary world. It raises questions such as: What are the embodied skills of doctors? How are these skills threatened or enhanced by technologies like the stethoscope? Should we defer medical judgements to a growing battery of technologies, or invest faith in the knowledge and skill of the physician? Are people happy to have consultations with computers and accept diagnoses made by algorithms? Is the presence and involvement of another human being important to medical interactions in some way? Though once an emblem of medical distance and detachment, in an increasingly technologically sophisticated world the stethoscope has come to stand for the doctor at the bedside, talking and *listening* to the patient. It represents a human and humane type of medicine, which some perceive to be under threat.

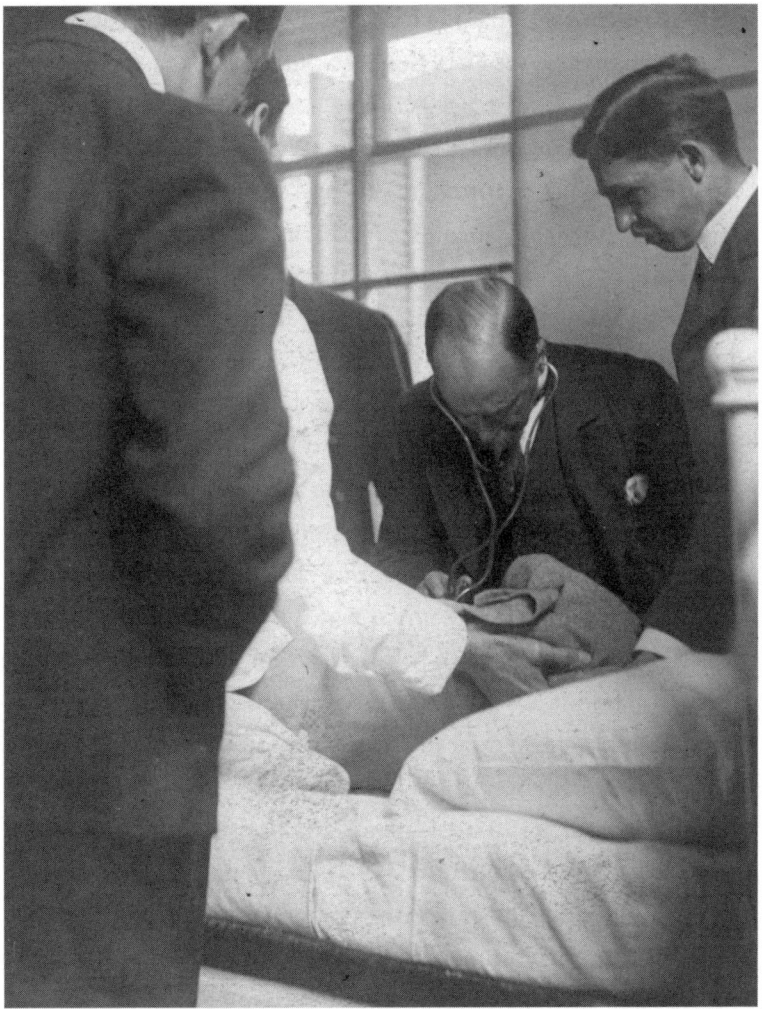

Canadian physician William Osler treating a patient,
showing the act of listening with a stethoscope, c. 1903.

Studying the cultural history of the stethoscope through a contemporary lens also offers insights into the ways in which bodies have been conceptualized and approached, whether that be the bodies of doctors, students or patients. The stethoscope is an important part of these shifting understandings. For instance, the invention of the stethoscope changed the way that doctors viewed both themselves and their patients, as they became less interpreters of the patient's words and more interrogators of the body. For medical students, learning to use a stethoscope became part of the essential identity work of doctoring, channelling skill and attention but also socializing them into their profession. Patients' engagements with stethoscopes over time also reveal changes in their capacity to countenance intrusive examinations and tolerate the physical proximity of the doctor. In the Victorian era, some women in particular were not prepared to expose themselves to stethoscopic examination, whereas today most patients regard it as routine.

Another reason to attend closely to the stethoscope is that it is important to understand how daily acts of listening in contemporary life have been shaped by historical developments. Headphone or earphone listening, for example, which is now ubiquitous, allows the listener to focus on sounds isolated from other sensory experiences. The sounds have been hived off from the soundscape more generally, and the listener can concentrate on them in a privatized auditory space. The sound studies scholar Jonathan Sterne argues that the origin of this way of listening can be found in the stethoscope.[1] It was the first technology to allow sounds to be made the focus of an individual listener's experience, attention and thought. This way of approaching sound and structuring auditory space was then replicated in numerous subsequent technologies: the telephone, headsets for gramophones, dictation machines and so on. We find it continued today in the auditory isolation brought by the use of headphones and earphones, and in this sense, stethoscopic listening is more prevalent today than it has ever been.

In this book we invite readers to tune in to the various histories and stories of the deceptively simple stethoscope. We have searched for sources which do not reiterate the common white male history of medicine, or the mainstream biomedical narrative, yet it is these accounts which are perpetuated in the vast majority of historical texts and medical journal articles, as well as in news and magazine articles about the stethoscope. We do not propose a radical alternative here, but we try when possible to draw on studies from outside Europe and North America and to bring in perspectives about uses of the stethoscope in, for instance, nursing, midwifery, engineering and veterinary practice too. We also include instances of the stethoscope's use in art and literature to show the variety of ways in which the instrument has been used and thought about.

In the opening chapter, we explore narratives of the invention of the stethoscope in the early nineteenth century and the context within which that invention is thought to have occurred. We situate stethoscopic listening as a technique that arose from other sensory and especially auditory methods for examining the body that were already in use in medicine at the time. We then move on, in the second chapter, to explore the reception of the stethoscope and the circumstances that conspired to facilitate and impede its acceptance in centres of medical learning across Europe and beyond. Although the stethoscope was eventually adopted widely, it was not used in quite the same way everywhere and was often given a distinctive local twist.

As use of the stethoscope spread around the world, it also spread around the body. We might find it easiest to imagine a doctor listening to a patient's chest, but in Chapter Three we see that the stethoscope has also been used, for instance, on the eyes, stomach, pubis and sacrum. Indeed, there are few places on the body that the instrument has not reached. This chapter also explores the profound effects that stethoscopic listening had on the doctor–patient relationship in the nineteenth century, as patients became obliged to make their bodies available for physical examination. At the same time, the technique also placed demands on the bodies of doctors, who had to master new sensory and interpersonal skills. Chapter Four shows how use of the stethoscope became routine in medical practice, so that by the twentieth century the instrument's status as a symbol of the medical profession was confirmed. But we also see that stethoscopic listening came to be used by people working in a variety of fields outside human medicine, such as engineering, veterinary work and in military activity. The chapter highlights the stethoscope's versatility and relevance to a wide range of situations and contexts.

Stethoscopic listening has been and continues to be a difficult skill to master, and in Chapter Five we focus on the various strategies that teachers have used to align their own sonic expertise with those of their students. We bring to the fore the particular places, techniques and technologies involved in transferring knowledge of how to handle the stethoscope and interpret sounds across generations of doctors. The question always lingers, however, as to how relevant it is to teach stethoscopic listening in contemporary medical contexts where other technologies that are often regarded as faster and more objective are in operation. This brings us, in Chapter Six, to look at threats to the stethoscope and claims that the use of the instrument is 'dying'. While technological innovation, especially in medical imaging, has affected the ways in which stethoscopic listening is used, and continues to make some doctors less dependent on the stethoscope, many argue that it is vital to preserve both the clinical skills and the human contact which use of the instrument requires. At the same time, in parts of the world where access to expensive equipment is limited, the stethoscope remains a core diagnostic technology.

The stethoscope was created through a process of tinkering, and the instrument has undergone continual modification and adaptation. Chapter Seven considers the ways in which the stethoscope has been improvised with and integrated into various medical systems around the world, as well as different ways in which it has been 'hacked'. We chart a history of innovation, through electrification to the digitization of stethoscopic listening, and show how current technologies such as 3D printing are working to complement rather than compete with the stethoscope. We bring these discussions into the present, a world which has been affected by a pandemic and the growing presence of artificial intelligence and machine learning in medical diagnosis.

Despite our focus on human healthcare, throughout this book we show how over the last two centuries stethoscopes have also lent themselves to all kinds of imaginative renderings by novelists and poets. Artists have experimented with the stethoscope as a metaphor and an interface in installations, and film-makers have used the device to gesture towards presences that are close to but beyond our direct visual perception. We demonstrate the stethoscope's remarkable capacity to stimulate thought and feeling. Thus, the book extends its gaze to all who engage with the stethoscope, and charts the variety of roles this iconic, polydimensional and polysemic instrument has played over time. Combining historical context, fictional stories and contemporary insights, we show that the stethoscope is never just one thing, and nor are the bodies it encounters. The stethoscope takes a vast number of forms, serves an astonishing range of purposes and is open to continual reinterpretation. The stethoscope is plural, and this is the key to its enduring presence and popularity.

We, the authors of this book, have of course been fascinated by the stethoscope ourselves. As anthropologists we have both carried out ethnographic fieldwork focusing on the instrument's use. Tom spent a year as a researcher among 'the white-coated tribe' at St Thomas' Hospital in London, where he accompanied groups of medical students as they embarked on their clinical training.[2] He held the position of Honorary Observer, which meant that he could sit in on a wide range of different interactions. After spending time with doctors and patients, he wanted to know more about how the skill of stethoscopic listening was passed on to novices, and so became a novice himself, attending the classes at which medical students learned to listen. During this apprenticeship, Tom learned about the skill of using the stethoscope through his own hands- (and ears-) on experience. He focused predominantly on listening to heart sounds.

Anna came to her research some years later from a different perspective. She had already trained and worked as a doctor but had chosen to move away from practising medicine to study it as a social scientist. In this capacity, she embarked on a project that focused on the use of the stethoscope in listening to lung sounds and on the ways in which medical students were taught to identify and interpret these diagnostic signs.[3] Digging out her old stethoscope, which had been hidden away in the dressing-up box at her parents' home, she embedded herself in classes at medical schools in Melbourne, Australia, and Maastricht, in the Netherlands. These medical schools were different from those in which she had trained, which helped her to see and hear the resources that the students used afresh.

We have both, then, had direct access into the closed world of medical learning in our studies. We make use of our notes and recollections from this fieldwork, and have reached out to those we have met and studied. But like many others, we have also encountered the stethoscope as patients, and as parents of patients.

At one point in his early twenties, Tom developed a nasty cough and went to see his General Practitioner (known in some parts of the world as a family doctor). The doctor listened to his chest with a stethoscope, pausing to listen to his heart as well as his lungs. He remembers feeling that the doctor seemed to be taking a considerable amount of time over it. There was a difficult moment of silence and waiting. 'Ah yes,' the doctor said presently, 'you most definitely have a heart murmur.' Tom was sent for an echocardiogram. It turned out that the murmur was very slight. It had nothing to do with his cough and did not pose a threat to his general health. But he nonetheless felt a complex mixture of emotions. There was surprise, and a sense of exposure and vulnerability when faced with the doctor's auditory gaze. He wondered how the doctor could possibly know his body better than he did himself. Part of him even felt he would have preferred not to know what the doctor had heard. At the same time, he was intrigued by the way the sound in his heart had been detected and interpreted.

Shortly after Anna's son was born, a doctor checked him over and listened to his heart and lungs. Anna could tell immediately that the clinician had heard something because she cocked her head slightly to one side and kept the stethoscope firmly in place with her fingers, seeming to listen more intently than before. Anna and her husband held their breaths. Blissfully unaware of the possibilities of what the doctor might find, their baby son played with the black plastic tubing of the stethoscope, wiggling it back and forth with his tiny fingers. Several doctors later, Anna and her family found themselves waiting to see the cardiologist. He jumped up almost immediately, stethoscope in hand, to listen to the baby's heart. Before long they were moved to an echocardiogram machine, and a judgement was finally made that the murmur that had been detected did not signify anything serious. There was a sigh of relief.

We describe these experiences of being on the 'receiving end' of the stethoscope because they exemplify the position from which so many people who are not medical professionals also experience it. These perspectives, and the strong emotions which often accompany them, are important to keep in mind. They add to and enhance our understanding of an object otherwise so closely linked to doctors and so richly storied by them.

1

Revelation

So where to begin our stethoscope story? In this chapter we consider the circumstances in which the first stethoscope came into being. The person widely credited with the invention of the stethoscope is the elegantly named René-Théophile-Hyacinthe Laennec (1781–1826). Born in Brittany in the northwest of France, his mother died when he was five and he was raised largely by his physician uncle, Guillaume-François Laennec. In a life marked by the political turmoil of the French Revolution, at the age of fourteen, with the help of Guillaume, Laennec enlisted as medical aide (surgeon third class) in the revolutionary army. By the age of sixteen he was committed to becoming a doctor.[1] After his stint in the army, Laennec returned to school and devoted himself to studying medical books, his uncle noting that he showed great promise as a physician. Laennec's father, however, was reluctant to provide the financial assistance necessary for him to complete his education. Always rather delicate, Laennec suffered serious illness and, following his recovery, briefly enlisted in the army again before his father finally provided funds for him to study medicine.[2] Laennec arrived in Paris and enrolled at the medical school in 1801, at the age of twenty.

At the time of Laennec's arrival, Paris was a major European centre of medical education, but importantly, it was also a centre for an emerging way of thinking about and researching disease. As Jacalyn Duffin, one of Laennec's biographers, writes: 'In the late eighteenth century, diseases were classified on the basis of symptoms, that is, the subjective feelings of sickness experienced by patients and "examined" by doctors through interview and observation.'[3] Nosologists, those who studied diseases, organized them into types based on symptoms. Doctors gave a diagnosis, a prognosis and treatment based on what they perceived to be the type and character of the symptoms a patient displayed (their order, intensity, pattern and so on). While they did examine patients, using their senses to detect characteristic physical signs they might associate with particular diseases, it was the patient's account of the symptoms that received most attention.

Although autopsy and dissection had long been a part of efforts to understand disease in Western medicine, in the Paris of the late eighteenth and early nineteenth centuries, research-focused physicians linked to the city's most prestigious medical institutions were increasingly focused on what became known as 'pathological anatomy'. By opening up corpses and making close observations of the internal organs and tissues, they sought to identify the organic traces of disease that were otherwise hidden from the physician. This represented a new epistemology of disease, that is, a particular strategy for investigating but also understanding it. Disease did not just equate to the symptoms suffered by the patient, but was related to underlying organic changes of which patients themselves were generally unaware.

Laennec, then, arrived into an environment in which pathological anatomy was emerging as a new science. Marie-François Xavier Bichat (1771–1802), for example, who is associated perhaps more than anyone else with pathological anatomy, worked in Paris and had published several important works in the years preceding Laennec's arrival. Bichat died in 1802 at the age of thirty, but his work had a major impact on the Parisian medical scene. Laennec's friend and later collaborator Gaspard-Laurent Bayle (1774–1816) wrote a thesis (which he defended shortly before Bichat's death) that proposed a classification of disease based not on the symptoms, which he argued were too many and too varied to be reliable indicators of particular diseases, but on the site and character of underlying organic lesions discovered during autopsy. In due course, Laennec would make important contributions to pathological anatomy himself.

Another key figure in Parisian medicine at this time was Jean-Nicolas Corvisart (1755–1821), a professor of medicine and one of Laennec's teachers. He was renowned for his remarkably acute powers of observation, diagnostic skill and for the accuracy of his prognoses. For some time he served as personal physician to Napoleon Bonaparte, and in Paris today there is a street, rue de Corvisart, and a Metro station named after him. It is said that Corvisart once examined a portrait and exclaimed 'If the painter were exact, the original died from heart disease.' He was correct. He was also apparently sometimes able to make a diagnosis while some distance away from a patient.[4]

Corvisart advocated the value of clinical skills, and one of those he practised himself was percussion. This technique of 'listen-touch' involves striking a finger of one hand with a finger of the other on particular sites of the body in order to make the spaces that are formed or enclosed by tissues or organs resonate. Corvisart had appropriated percussion from the Austrian physician Josef Leopold Auenbrügger (1722–1809). Auenbrügger was the son of an inn-keeper, and as a boy had learned to test the fullness of barrels by knocking on them. Later in life he substituted beer kegs for rib cages, and noted that if struck with a finger, healthy and unhealthy chests sounded differently: 'a healthy chest sounded like a cloth-covered drum; by contrast, a muffled sound or one of high pitch indicated pulmonary disease.'[5] Corvisart had encountered (and later translated) Auenbrügger's treatise on percussion and took up the technique in his clinical practice and research. By ensuring that the deaths of patients whom he had examined were investigated at autopsy, Corvisart was able to build up a knowledge of how percussion sounds and other signs related to anatomical changes, especially in the heart. Indeed, he specialized in the study of heart disease and is widely regarded as the founder of modern cardiology.

In percussion (a technique that has become routine in modern medical practice), sounds generated by the physician through tapping are positioned as valuable diagnostic signs which in some cases can indicate underlying anatomical change. We see this principle reproduced in stethoscopic listening. Laennec also learned from Corvisart the value of sharp clinical skills, including attention to sounds. For instance, he noted from Corvisart's lectures that in some cases heart palpitations could be so intense that the heart could be heard beating

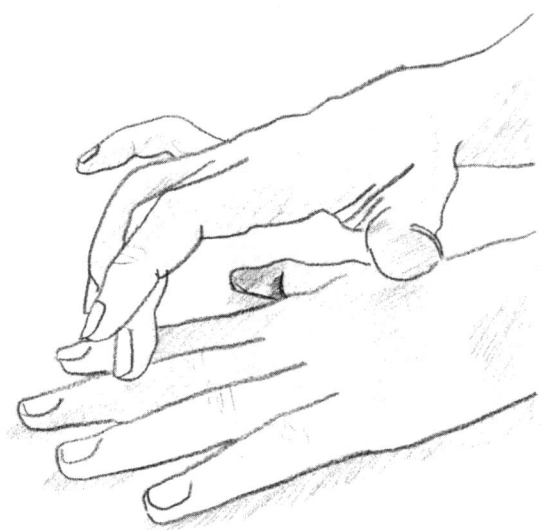

Illustration of hands performing percussion.

against the chest wall.[6] Significantly, percussion is also a technique of 'listen-touch', and we might understand stethoscopic listening, too, to be a form of auditory tactility; the instrument touches the body while the doctor listens.

During his studies, Laennec published a number of articles and reports detailing findings in anatomy and pathology, acquiring some recognition as a researcher. He also won prizes in surgery and medicine and wrote a thesis on Hippocrates which showed his familiarity with classical medical ideas. On his graduation he was confident that he would soon receive a good position in the Paris medical school, but such an opportunity did not materialize for some time. Instead Laennec continued to conduct research, writing and editing articles and reviews. He also took on paying patients to ease his financial difficulties (the academic work was not lucrative). His practice

Jean-Nicolas Corvisart, 19th century, lithograph by Bornemann after F.P.S. Gérard.

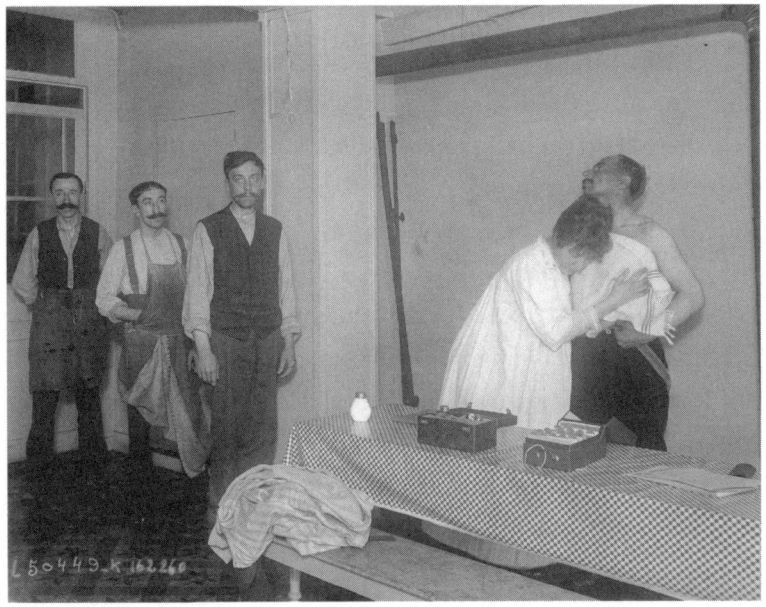

Doctor examining a patient, press photograph, 1917.

expanded and he became a well-known physician with a number
of clients from the political and religious elite. In September 1816
Laennec was appointed doctor at Paris's Necker Hospital, a one-
hundred-bed institution which accepted male and female patients.
This appointment was key. Laennec was already an expert pathol-
ogist with highly developed clinical skills, but the hospital gave
him ready access to large numbers of patients whom he could visit
and revisit. Crucially, it also gave him access to the autopsy facilities
needed to conduct the research that, as we will see, underpinned
stethoscopic listening. Without these resources, Laennec would
have been unable to embark on his stethoscopic work.[7]

The Invention

Understanding the medical context within which Laennec studied helps us to grasp the background against which the stethoscope came into existence, but various other influences may also have factored into the discovery. For instance, Laennec was familiar with work by the English physician William Hyde Wollaston (1766–1828), who in 1810 conducted a series of experiments in which he used a long 'stick' to transmit the sounds made by muscles from his foot to his ear. It may also be that Laennec was adapting acoustic experiments that he had learned about at school, or read about in the work of the Roman scientist Pliny the Elder, who tested the capacity of wood to conduct sound. But although Laennec does refer to Pliny in his writing on other topics, he does not do so in relation to his invention of the stethoscope.

We know that Laennec had also read the thesis of one of his contemporaries at medical school, Matthieu-François-Régis Buisson (1776–1804). Writing about the functions of the human organism, Buisson drew a distinction between the active and passive aspects of its various capacities, including hearing. He referred to passive hearing as 'audition', while active listening was 'auscultation' – from the Latin verb *auscultare*, meaning 'to listen'. Buisson also referred to different discernible qualities or types of human voice, and these came to be of great relevance for Laennec in his early experiments with the stethoscope. In addition, while a student, Laennec had written his final thesis on aspects of Hippocratic medicine. Hippocratic literature (dating from around 410 BC) makes reference to a practice that became known as 'immediate auscultation', which involved applying the ear directly to a patient's body. This could be combined with 'succussion', or shaking a patient, which could produce a splashing sound (now known as the

Blind doctor Albert-André Nast using his ear instead of a stethoscope, Chelles, 1953, photograph by Thomas D. McAvoy.

Hippocratic succussion splash) if fluid was present inside the chest cavity. Though rarely described and hence probably infrequently practised, Laennec knew about immediate auscultation, and some doctors still occasionally listen in this way today. The stethoscope, then, emerged in a context where numerous relevant ideas and practices were circulating.[8]

The popular story or legend surrounding the invention of the stethoscope is that one day in 1816, Laennec was walking through the courtyard of the Louvre near the Necker hospital. He was wondering what to do about a particular patient of his, a young woman with a heart complaint. He had been unable to learn much from her own accounts of her disease. At the same time, Laennec felt that pressing an ear to the chest of a young woman, as was required for immediate auscultation, would be improper, and he had been unable to discern anything through palpation or percussion.[9] At that moment Laennec happened to see some children playing a game with a log that was resting on top of a pile of builders' rubble. Those at one end were pressing their ears to the wood, and were able to hear knocks and scratches made by their playmates at the other end. This gave Laennec his 'Eureka!' moment. Returning to his patient's bedside, he rolled a stack of paper into a cylinder so that it resembled a log and pressed it to her heart. He found he could hear her heartbeat clearly.

This story appears in various accounts of the stethoscope's invention, including the one in the film based on Laennec's life: *Docteur Laennec* (1949). It is a sort of creation myth; an apocryphal tale. The detail of Laennec being inspired by children playing with a log might originate from an associate of Laennec's named Jacques-Alexandre Le Jumeau de Kergaradec (c. 1788–1877), who claimed to have heard the story directly from Laennec himself.[10] Laennec, however, does not mention either the garden of the Louvre or the children's acoustic game in the account of the invention that he gives in his *Treatise on the Diseases of the Chest and on Mediate Auscultation* (though editions and translations vary):

In 1816, I was consulted by a young woman labouring under the general symptoms of diseased heart, and in whose case percussion and the application of the hand were of little avail on account of the great degree of fatness. The other method just mentioned [he is referring to immediate auscultation here] being rendered inadmissible by the age and sex of the patient, I happened to recollect a simple and well-known fact in acoustics, and fancied, at the same time, that it might be turned to some use on the present occasion. The fact I allude to is the augmented impression of sound when conveyed through certain bodies, – as when we hear the scratch of a pin at one end of a piece of wood, on applying our ear to the other. Immediately, on this suggestion, I rolled a quire of paper [25 sheets] into a kind of cylinder and applied one end of it to the region of the heart and the other to my ear, and was not a little surprised and pleased to find that I could thereby perceive the action of the heart in a manner much more clear and distinct than I had ever been able to do by the immediate application of the ear.[11]

There is much that this account does not tell us about the precise sequence of happenings, thoughts and interactions that surrounded what we now call the invention of the stethoscope. Laennec's own version of the story is a construction, a post-hoc account written 'with full knowledge of the ultimate significance of the event'.[12] It may also have been deliberately crafted by Laennec in such a way as to disguise his influences and protect his claim to be the pioneer of 'mediate auscultation', that is, auscultation mediated by an instrument.

The identity of the first person to be auscultated with Laennec's stethoscope is also unknown. Some evidence suggests it could have been Jacquemine Guichard Argou (1779–1847), the woman Laennec went on to marry in 1824, but this is uncertain.[13] The sources all agree, however, that the first stethoscope was a simple cylinder of paper. Indeed, 'the stethoscope' did not actually acquire that name until 1818. Laennec considered his first improvised device to be so basic as not to require a name. He referred to his paper cylinder as just that: *le cylindre*.

Development

Laennec soon began to try out his cylinder on other patients, and found it allowed him to hear a variety of heart and lung sounds, many of which it had not been possible to detect before. Quite soon after the initial discovery, he began a programme of research using the cylinder. His method involved making links between clinical signs observed at the bedside and anatomical changes found at autopsy. Along with other forms of examination, Laennec would listen to the chests of his patients with the cylinder, making careful notes on what sounds he heard and where on the patient's body he placed the cylinder in order to detect them. He would repeat his examinations and listenings as his patients' diseases progressed. When those patients died he took a keen interest in the autopsy findings (he conducted numerous autopsies himself). In this way he was able to begin to link the sounds he heard to underlying pathologies. As well as producing a new technology, then, Laennec had also created a new framework for the interpretation of the sounds heard through it. They became meaningful through their connection with emerging concepts of disease generated by pathological anatomy.

One important early finding for Laennec was that if he listened to their chests *while they spoke*, some of his patients had particular points where their voice seemed to be transmitted to his ear especially directly and strongly. He called this particular acoustic phenomenon 'pectoriloquy' (also translated as 'pectoriloquism'), which means 'the chest speaks', in the same way that ventriloquy means 'the stomach speaks'.[14] Autopsies of those who subsequently died revealed cavities in the lung in exactly the place from which the pectoriloquy had seemed to originate. Laennec recognized that these cavities were caused by what was then known as 'phthisis' but is now called 'pulmonary tuberculosis'. He concluded that pectoriloquy was a reliable sign that excavations in the lung, and hence tuberculosis, were present. This was a major advance, as it had not previously been possible to ascertain the presence of lung cavities in a living patient. Laennec concluded that pectoriloquy was a pathognomonic sign of tuberculosis, meaning it was a sign that reliably indicated this specific disease. Patients arriving at a clinic with the condition might have symptoms such as fever or catarrh, but if this acoustic sign was found it would now be possible to give a definite diagnosis of a cavity in the lung and hence of phthisis as the underlying disease.

Pectoriloquy was perhaps Laennec's most important stethoscopic discovery, but he also identified other diagnostic signs. For instance, the cylinder allowed him to notice that when he listened to some of his patients' voices through the chest wall it sounded 'as if a kind of silvery voice, of a sharper and shriller tone than that of the patient, was vibrating on the surface of the lungs, sounding more like the echo of the voice than the voice itself'.[15] At the same time, the voice sometimes took on 'a trembling or bleating sound, like the voice of a goat'.[16] Laennec called this vocal quality 'aegophonism' or 'egophony', and post-mortem findings enabled him to associate it with pleural effusion, sometimes known as 'water on the lungs', where fluid builds up between the thin membranes lining the lungs. He also noticed diminished breath sounds in patients suffering from emphysema, a sign which he was able to link to changes in the lung tissue found at autopsy. Laennec was surprised by how frequently he detected these reduced breath sounds, which made him realize that emphysema was far more common than was generally thought at the time – another significant medical advance.

Laennec also listened to the heart, an organ which had been a specialism of his teacher, Corvisart, and which was of great interest to other anatomists and physicians. Laennec heard and described a wide variety of heart sounds in his patients, including the normal sounds of the heartbeat (the so-called 'lubb-dupp') and several further variations. He also heard heart murmurs, which today we know are caused when one (or potentially more) of the heart valves becomes diseased, producing turbulence in the blood flow. This turbulence generates an audible vibration which in rare cases is loud and can be detected by the doctor (and even the patient) with the naked ear, but is usually very soft, so that an amplifying device is required in order to hear it. Sometimes murmurs are audible in other blood vessels when the blood flow is turbulent. Laennec noticed murmurs of different types and intensities, describing their sonic characteristics by likening them to the sounds of everyday objects such as bellows, wood files and saws. In one famous case he used a stave with a treble clef, notes and slurs to document a sound he had heard in the carotid artery of a female patient.[17]

Laennec's findings on the auscultation of the heart have generally been regarded as less valuable than those relating to the lungs.

Laennec's musical notation documenting the sounds
he heard in the carotid artery of a woman who
consulted him in March 1824.

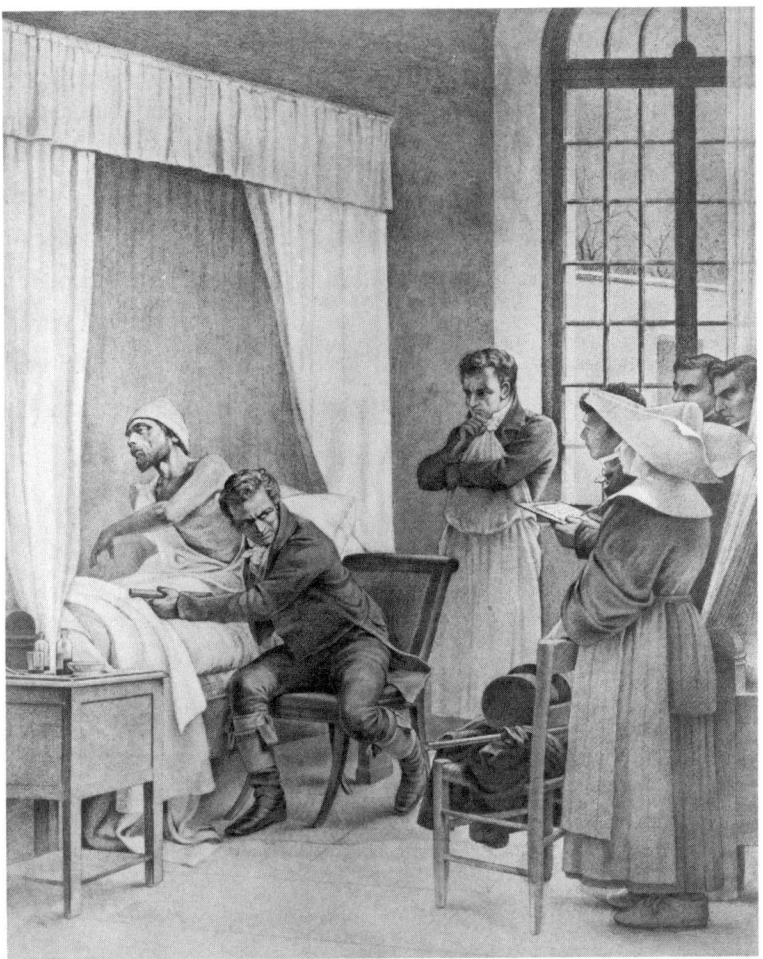

Laennec auscultating a tubercular patient at the Necker Hospital, Paris, 1816, heliogravure after a painting by Théobald Chartran.

This is mainly because his ideas did not align with what later became accepted knowledge about the way the heart functions, and hence the way heart sounds are produced. Laennec believed that the normal heart sounds were generated by the muscular contractions of the atria and ventricles, rather than, as later became accepted, the closure of the heart valves. He also tended to regard heart murmurs as the result of muscular spasms rather than of abnormalities in the valves of the heart, a theory that has since been rejected. But Laennec did identify a number of signs that have stood the test of time. For instance, he noticed that there was an inverse relationship between the loudness of the heart sounds and the heart's muscle mass: loud heart sounds implied thin muscle walls with enlarged chambers, while quiet or absent ones indicated thickened muscles. So although Laennec's ideas as to the cause of some sounds may have been rejected, he was able to identify sounds which went on to become recognized clinical signs, and which were later made the subject of further investigation by his students, colleagues, supporters and detractors.

The Emergence of 'the Stethoscope'

Laennec describes his original paper cylinder as having been 'formed of three quires, compactly rolled together and kept in shape by paste'.[18] Soon after his initial improvisation with a paper tube, Laennec acquired a lathe, taught himself how to operate it and began to make cylinders from other materials. He found those of high density (such as glass or metal) and low density (he mentions a thin material of beaten gold known as 'goldbeater's skin') to be ineffective at conveying sounds, while glass and metal had the added disadvantages of being heavy and cold in winter. He found paper and light wood to be more suitable, and settled on a design which consisted of

> a cylinder of wood an inch and a half in diameter and a foot long, perforated longitudinally by a bore three lines wide [three 'lines' equates to 6 mm, just under a quarter of an inch], and hollowed out into a funnel-shape, to the depth of an inch and a half at one of its extremities.[19]

Partly for portability, and partly to allow it to be used at half-length if desired, the cylinder could be divided into two sections. The funnel-shaped end, to Laennec, was especially useful in exploring the varied pitches of the sounds of respiration, but he found that when examining the heart and voice a simple tube was better; he added a stopper or plug that had itself been drilled, so that when fitted in the funnel-shaped end it gave the instrument's central bore a consistent diameter.[20] These modifications anticipated the addition of the diaphragm (for higher-pitched sounds) and the bell (for lower-pitched ones) in later designs of stethoscope – though it might equally be the case that later modifications simply resurrected Laennec's original ones.

Those who observed Laennec using his device and who knew of the discoveries he had made suggested a variety of names for it, including the *sonometre*, *pectoriloque*, *pectoriloquie*, *thoraciloque* and *cornet medicale*, but Laennec eventually settled on 'stethoscope', a combination of the Greek words for 'chest' (*stethos*) and 'to examine' or 'explore' (*skopein*).[21] There are still a few stethoscopes of Laennec's own making in existence, and many which conform to his original design, for example the one shown below, which is held at the Wellcome Collection in London. Laennec continued to experiment with stethoscope designs until his death in 1826.

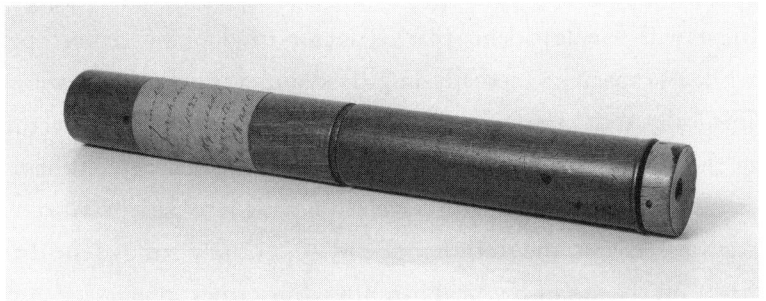

Monaural stethoscope, Laennec type.

Several people have remarked to us that tubular monaural stethoscopes such as this specimen look rather like musical instruments. In fact, as with Auenbrügger, the originator of percussion, Laennec was musical. He played the flute, a skill which may have helped him both to make stethoscopes and to handle them adeptly. His musical ear would also have been useful in enabling him to make fine distinctions between the different sounds that he heard in the chests of his patients. As we will see later, however, while Laennec's own aptitude for making subtle sonic distinctions may have been helpful to him in developing the technique, it later acted as a barrier to the ready acceptance of mediate auscultation by some other doctors; not all were, or are, as golden-eared as Laennec.

As indicated above, Laennec knew about immediate auscultation, the technique of pressing an ear to a patient's chest in order to listen, but he came to consider it unsatisfactory in important ways. For instance, he wrote that 'in that class of persons found in hospitals it is disgusting.'[22] He argued that the sick poor were often dirty or bathed in perspiration, so that pressing an ear to their chests was unpleasant. The physical intimacy involved in immediate auscultation also meant it was 'not only indelicate but often impracticable' when patients were female.[23] In fact, some believe that it was primarily the issues of hygiene involved in immediate auscultation, and also the difficulty of male physicians managing intimate engagements with female patients, that led doctors to adopt the stethoscope, while its capacity to actually amplify sound was open to question. Ironically, today the stethoscope is known to be a potential vector in the spread of germs in hospitals, as microbes can attach themselves to the instrument and travel from patient to patient. Also, as we will see later, the stethoscope did not entirely remove the difficult intimacies of medical listening, but rather moderated and changed them.

Laennec knew, too, of percussion as practised and taught by Corvisart. It was, he wrote, 'one of the most valuable discoveries ever made in medicine', but he found the technique to be 'very incomplete' because it essentially gave only an indication of fullness and emptiness.[24] Particularly in relation to diseases of the heart, he suggested, 'we regret the insufficiency of this method, and wish for something more precise.'[25] Introducing stethoscopic listening, however, he advocated that it be used in conjunction with percussion, as well as succussion and immediate auscultation, 'all of which methods,' he wrote, 'often useless in themselves, become of great value when combined with the results procured through the medium of the Stethoscope'.[26]

The invention of the stethoscope would make Laennec famous both nationally and internationally in his own lifetime. But while Laennec will be present in later chapters, it is interesting to look ahead to the end of his life; if he lived and rose to fame by the stethoscope, there is perhaps a sense in which he died by it too. Laennec did not enjoy good health and is recognized to have been a consumptive; in other words, he suffered from what is now known as tuberculosis, as had so many of his patients (and he had, of course, identified a key diagnostic sign of this same disease in the form of the pectoriloquy). The theory of pathogenicity – the now well-established idea that entities such as bacteria, viruses, fungi and parasites could cause disease – and the notion that phthisis could be contracted through contact with the body fluids of an infected person, was not accepted during Laennec's lifetime, but he lived and worked in an environment in which the disease was endemic. He also recognized that the demands of his medical work compromised his health, and this was no doubt true of his stethoscopic labours.

Like many of us, Laennec was a close student of his own health. During his lifetime he observed numerous symptoms and made a variety of self-diagnoses. He would not have been able to apply the stethoscope to himself (that would have required a flexible device, and these had not yet been developed), but near the end of his life he allowed himself to be examined by several colleagues.[27] He did not spare himself close stethoscopic scrutiny, and there is perhaps a certain equality or equilibrium of listening here. Laennec also recognized that his legacy to medicine was closely bound up with the stethoscope and the ongoing practice of mediate auscultation, as was at least one of his important personal relationships. Laennec was close to his younger cousin, Mériadec, who had followed him into medicine and whose career Laennec had closely mentored. Mériadec had written his thesis on mediate auscultation and was a strong advocate of the technique. He was one of the people that Laennec allowed to listen to his chest near the end of his life. In a letter the dying Laennec, who had no children of his own, declared to Mériadec (in what could be considered an act of patriarchal legacy that permeates the history of the stethoscope): 'you are my spiritual son and the heir of my stethoscope.'[28]

Implications

The practice of stethoscopic listening emerged during a time of profound change in the medical profession. Doctors, especially those in the Paris medical school, were increasingly focused on pathological anatomy, and Laennec was deeply involved in this work and the ideas about disease which developed through it. The stethoscope was conceived as a way of using the sense of hearing to see what was happening inside the body. It allowed an acoustic illumination of the bodily interior, enabling Laennec to make informed judgements as to the absence, presence and stage of particular anatomical changes.

The philosopher, historian and social theorist Michel Foucault refers to the medical 'gaze' that he considers to have emerged in the late eighteenth and early nineteenth centuries, and which he associates directly with the project of pathological anatomy in French medicine. The gaze refers to a skilled form of attention that is brought into play in the examination of the patient at the bedside but also in the close scrutiny of the internal organs at autopsy.[29] The term 'gaze', however, refers to more than solely visual attention. According to Foucault it 'embraces more than is said by the word "gaze" alone'.[30] It contains within a single structure different sensorial fields; it is a gaze 'that touches, hears, and moreover and not by essence or necessity, sees'.[31] The stethoscope is both subsumed within and contributes to the power of the clinical gaze, specifically in its capacity 'to recognise on the living body what was revealed on the corpse by dissection'.[32] It enables the user to 'look' inside and essentially anatomize a living patient. In fact, the technique actually required an animate body, because only then would there be the breath, voice and so on necessary for investigative listening to be beneficial. Mediate auscultation facilitated the 'autopsy of the living'.[33]

The anatomical turn, of which the stethoscope can be understood as a product, meant the doctor was becoming less an interpreter of the patient's words and more an interrogator of the body, forming clinical judgements based on an independent assessment of diagnostic signs. 'In Laennec's lifetime,' writes Duffin, 'disease concepts changed from constructs based on patients' subjective symptoms or feelings, described in the patient's history, to concepts based on specific changes in the patient's body, detected objectively through physical examination.'[34] In the nineteenth century the stethoscope would become emblematic of this important shift in medical thought and practice. The device made the body eloquent irrespective of its owner's capacity to articulate felt sensations. The body now described its condition through its sounds, and importantly, and at least in principle, those sounds didn't ramble, lie or mislead the way patients could. The stethoscope entailed a redirection of the doctor's auditory focus from the patient's words to his or her body. In actual fact, as described above, mediate auscultation still involved attending to the voice, but it was the qualities of the patient's voice rather than the meaning of his or her spoken words that became the primary focus of medical attention and interpretation.

Foucault describes the stethoscope as 'solidified distance' which 'transmits profound and invisible events along a semi-tactile, semi-auditory axis'.[35] For him, the instrument 'authorizes a withdrawal'.[36] It is the materialization of a step back from the patient, made partly due to doctors' disgust and partly to protect their perceived sense of decency and decorum. But the distance between doctor and patient also reflects a distance between knower and known. The stethoscope positioned the doctor as a figure who held skill and knowledge above and independently of the patient. Just as doctors no longer needed their patients to provide all the information required for diagnosis, mediate auscultation made it possible for them to recognize the presence of a disease without the patient necessarily even feeling unwell, breaking a long-established link between *feeling* ill and *being* ill, and hence *knowing* whether one was sick or not.[37] With stethoscopic listening, the doctor became the central arbiter over the distinction between health and disease.

2

Rise

How did use of the stethoscope begin to extend beyond Paris, eventually becoming widespread? Our minds might conjure up an animation sequence of stethoscopes moving across Europe and the rest of the world as if by their own volition, self-replicating and inserting themselves smoothly into the hands of waiting doctors. But seductive as this narrative might be, there is a great deal that it neglects or glosses over. Although by the end of the nineteenth century, according to Jonathan Sterne, 'stethoscopy was everywhere' and practitioners who did not use mediate auscultation risked their professional reputations, the technique, for various political and cultural reasons, was adopted more readily in some places than others.[1] In parts of Europe it was actively opposed by some doctors, while in India, for example, there was little such resistance.[2] At the same time, ways of using the stethoscope and interpreting the sounds heard through it sometimes underwent changes depending on local contexts. The instrument itself also proliferated into a variety of forms as doctors experimented with new designs, though such experimentation was an indulgence that some doctors serving the poor considered an unnecessary luxury.[3] In this chapter we unpack some of the complexities of the rise of the stethoscope.

The Practice Spreads

In 1819, Laennec formally introduced the stethoscope and his research into diseases of the chest in his treatise *De l'auscultation médiate* (On Mediate Auscultation). He offered stethoscopes of his own design for sale with the text for an extra 3 francs.[4] In the same year that the treatise was published it began to be reviewed in medical journals, initially in France but soon after in England and Germany. Reviews had been published as far afield as Spain and the United States by 1821. The book was initially perceived by many, first and foremost, as a contribution to pathological anatomy; the significance of the stethoscope and mediate auscultation was only realized more gradually. Before long, however, review articles were describing Laennec's stethoscope and the ways in which it could be used and applied, as well as the diagnostic significance of the sounds that Laennec claimed could be heard through it. The reviews meant that doctors across Europe and beyond started to learn about stethoscopic listening, some trying the technique for themselves. By 1822 Laennec learned that independent observers in Madrid, Edinburgh, Bonn, Berlin, Barcelona and Boston had confirmed some of his key findings.[5]

Laennec's treatise was translated into a variety of languages. John Forbes (1787–1861) produced four different English editions between 1821 and 1834, and some of these were reprinted in the United States, so that the text was available there by 1823. A German translation appeared in 1822 and a Belgian one in 1828, while a four-part Italian translation appeared between the years 1833 and 1836. This meant that a growing number of doctors, initially in Europe and then internationally, could gain a more detailed understanding of Laennec's ideas than had been possible through reviews alone. But mediate auscultation did not only travel through texts. Its spread was not just a matter of diffusing or disseminating information. It moved as a skill, a practice, carried in the bodies of those who learned and trained in the technique.

Paris in the first decades of the nineteenth century was a major centre for medical training, attracting students from far and wide. Laennec was also an energetic teacher. Numerous French students learned mediate auscultation from him, and he also taught a large number of international students who had heard about the stethoscope and were keen to receive instruction from the master himself. During his lifetime Laennec personally gave instruction on mediate auscultation to some three hundred foreign students.[6] They carried their learning with them to a variety of different countries, including England, Scotland, Germany, the Netherlands, Italy, Denmark, Norway, the United States and Canada, as well as India and countries in Latin America.[7]

Laennec kept records of his foreign students and sometimes commented on their skill. The performance, for instance, of Thomas Hodgkin (1798–1866), later a pioneer of mediate auscultation in England and the man responsible for introducing the stethoscope at Guy's Hospital in London, was recorded as 'assez bien'.[8] Charles James Blasius Williams (1805–1889), who later developed his own work on auscultation and was also an advocate of the technique, is recorded as having been 'très bien'.[9] John Elliotson (1791–1868), a friend of Charles Dickens, went to France for the purpose of learning to use the stethoscope from Laennec himself. Laennec, however, was too ill to attend the hospital and so Elliotson missed out on his instruction.[10] Despite his frustration at not having received personal tuition from the famous teacher, Elliotson later went on to become a strong proponent of the stethoscope in England.

So far we have painted a picture in which the spread of auscultation through a text-based system of knowledge transfer and through the practical skill of some early adopters appears to have been smooth and unproblematic. We have suggested that the stethoscope was quickly recognized as a useful and valuable innovation and was taken up with enthusiasm. We have also perhaps created the impression that as the technique began to develop outside Paris it remained consistent with an original version described by Laennec. This seamless story can be problematized by considering the differing ways in which mediate auscultation was received outside Paris. Local contexts affected the adoption of the stethoscope and sometimes changed the way in which it was used.

Edinburgh

Edinburgh was considered another important centre of medical research and training in the early nineteenth century.[11] A variety of factors helped to facilitate the rapid uptake of auscultation there. The same interest in pathological anatomy that was characteristic of Parisian medicine was also a feature of Scottish medicine. There was also a strong tradition of clinical training (that is, teaching at the bedside) in Edinburgh, so that mediate auscultation could be easily integrated into the existing system of medical education in Scotland.[12]

Practical trials of mediate auscultation were very promptly initiated in Edinburgh, and doctors there began experimenting with the stethoscope early in the 1820s.[13] By 1824, William Cullen (1798–1828), a grand-nephew of the more famous medical figure of the same name, was teaching the stethoscope in Edinburgh. He had studied under Laennec in Paris and gave a series of special lectures on the use of the instrument that were attended by students and fellow staff members. In 1826, James Crauford Gregory (1801–1832), who had also studied with Laennec, took up the position of physician to the Royal Infirmary. By the mid-1820s, then, alongside academic textual commentary and self-taught use, men trained by the inventor himself were teaching and demonstrating the use of the stethoscope in Edinburgh.

Cullen's teaching seems to have had considerable impact, not least on James Hope (1801–1841), who in 1825 submitted a thesis investigating the potential for the use of the stethoscope in the diagnosis of arterial disease. It included six case histories, all of which related to patients from the wards of the Edinburgh Royal Infirmary. Hope was even able to contradict and correct Laennec on an important point: Laennec had thought that mediate auscultation would not be of value in the diagnosis of aortic aneurysms, because the sounds associated with the aneurysms would be indistinguishable from the sounds of the heart itself. Hope showed that making a distinction between these sounds was in fact possible. He thought that Laennec had not realized this because he lacked experience in cases of aortic aneurysm. Other Edinburgh doctors also later challenged and developed Laennec's work. The existence of this research not only shows that by 1825 the stethoscope had come to have real value for Edinburgh's academic physicians, but it demonstrates their independence from Parisian work on mediate auscultation. Edinburgh doctors were making the technique their own.

By 1828 the stethoscope was being used routinely in clinical investigation and instruction in the Royal Infirmary.[14] The 1829 edition of the key medical textbook used in Edinburgh included revisions detailing the utility of the stethoscope in distinguishing different diseases of the chest. Changes to some classifications of diseases had also been made in line with insights gleaned from stethoscopic research. So while between the years 1820 and 1830 proponents of mediate auscultation had produced publications that actively sought to demonstrate the value of the stethoscope, by 1831 its utility no longer had to be 'argued for or displayed'.[15] Use of the instrument had become routine and unremarkable, if not in every case, at least in clinically interesting cases. 'Thus,' Malcolm Nicolson concludes, 'by 1831 the transfer of the innovation of stethoscopy from Paris to Edinburgh had been successfully accomplished.'[16] But though it may be difficult to find any evidence of resistance to the stethoscope in Edinburgh, many graduates who embraced the stethoscope did encounter opposition when they moved cities to practise medicine.

London

Laennec's treatise seems to have been generally well received by reviewers in the English capital. Some early, energetic proponents of mediate auscultation also lived and worked there. The fact that in 1826 *The Lancet* published its 'Directions for the Use of the Stethoscope' and made reference to a recent order for army surgeons to use the instrument and to report on their findings, with similar instructions issued to naval surgeons, indicates that the value of the stethoscope was recognized by some influential figures in London's medical community at that time.[17] The stethoscope was probably introduced to the armed forces by Sir James McGrigor (1771–1858), director general of the British Army Medical Services between 1815 and 1851. He too had made the trip to France to visit Laennec.[18] The introduction of the stethoscope to the British military initiated the use of the instrument across the British Empire.

The acceptance and adoption of mediate auscultation, however, seems to have been slower in London than in Edinburgh, and the technique met with resistance from some doctors. James Hope, mentioned above as the author of a thesis on the use of auscultation in the detection of aortic aneurysm, left Edinburgh in 1825 and toured the medical centres of Europe, receiving training in auscultation from Laennec's successor Auguste François Chomel (1788–1858) at the Hôpital de la Charité in Paris in 1826. He returned

Stethoscope being used by Surgeon-Captain Whiston in Sudan, drawing by H. M. Paget after a sketch by W. T. Maud, from *Graphic* (6 November 1897).

to London to begin practising. He found, however, that there was strong prejudice against auscultation in England and 'had to battle both mistrust and complacency in order to get the stethoscope accepted and widely used'.[19] John Elliotson, mentioned above as having been frustrated in his efforts to seek out instruction from Laennec, also found there was mistrust of the stethoscope and lack of information about the instrument in London's medical community. Scepticism as to its value continued during most of the 1830s, and even in the 1840s some London commentators bemoaned 'the existence of considerable ignorance of, and opposition to, the stethoscope among the metropolitan medical profession'.[20] The stethoscopists were 'a minority group, against whom there was a significant degree of prejudice'.[21]

Although the medical historian P. J. Bishop argues that the stethoscope was, broadly speaking, quickly integrated into medical practice in England, he writes that 'There were, of course, some who scoffed.'[22] He refers to a *Lancet* editorial of 1826 which mentions Dr Grant David Yeats's Croonian Lectures to the Royal College of Physicians of London. Yeats (1773–1836) apparently classed the stethoscope alongside acupuncture, metallic tractors (metal rods allegedly providing pain relief) and phrenology, declaring them all to be 'ephemeral follies'.[23] *The Lancet* also mentions some people maliciously referring to auscultators as 'Laennec trumpetters [*sic*]',[24] while Edward Bluth writes that during the 1820s and '30s 'the stethoscope was regarded by many as simply a new "gadget" and was not considered of any real clinical use in the great hospitals of London.'[25] In George Eliot's *Middlemarch*, published in 1871 but set between 1829 and 1831, Dr Tertius Lydgate examines Edward Casaubon with a stethoscope. In doing so he shows that he is a man who embraces modern medical methods, but he also irritates his more conservative colleagues.[26]

What was it, then, that lay behind a reluctance to accept the stethoscope in England? Historians have suggested a number of factors. One is that the English were less receptive to developments in French medicine than their Scottish counterparts. The Napoleonic wars had taken place between 1803 and 1815, and comments on the stethoscope by some London doctors reflected Francophobic sentiments. The instrument was referred to as a 'French bauble' and an 'insult to John Bull' (John Bull was the equivalent to Uncle Sam in the United States, a national personification of Britain and, in particular, England).[27] John Forbes, the translator of Laennec's treatise, was apparently convinced that it would require the publication of relevant English case studies for the technique of mediate auscultation to be accepted by an English audience.[28]

The prevailing academic and institutional context may also have affected the uptake of auscultation in London. In Edinburgh there was only one dominant medical school, meaning that the character and content of medical training could be determined by a relatively small group of academic physicians. London, however, had numerous teaching hospitals and private medical schools, none of which exerted any particular influence over the others. This made the achievement of consensus on the value of any innovation far harder in London than it was in Edinburgh.[29] In addition, in 1820s London lectures had become established as the key medical teaching technique. While they were effective for the communication of certain types of knowledge, they were not very suitable for the teaching of auscultation, which required practical instruction at the bedside.[30] Pedagogic techniques, then, did not provide the best conditions for mediate auscultation to develop. Indeed, one of the reasons medical students in the 1820s were so keen to go to Paris was that they were given 'free run of the public hospitals; it was easier to get ample bedside and dissection experience there than anywhere else.'[31] It was not until the 1830s and '40s that bedside teaching became more commonplace in England, creating opportunities for auscultation to flourish.

Vienna

Despite the availability of a German translation of Laennec's treatise and the fact that some students from German-speaking countries had received training in Paris, few physicians from these places were using the stethoscope even a decade after the instrument's invention. During the 1820s 'German physicians still relied almost exclusively on the interpretation of the patient's narrative or on ancient forms of rudimentary clinical examination.'[32] They conceived of illness as not simply a localized pathological change but 'a disturbance of the whole state of the body and of its relationship to its social and natural environment'.[33] A lengthy verbal consultation with the patient was held to be the best approach to diagnosis. There were, however, a few advocates of the Parisian tradition of pathological anatomy working in German-speaking countries at this time, and one of them was Joseph Škoda (1805–1881).

Škoda was a blacksmith's son who had walked from his native Pilsen in what was then Bohemia to study medicine in Vienna. He collaborated closely with Carl von Rokitansky (1804–1878), a committed advocate of pathological anatomy, who is said to have conducted more than 60,000 autopsies during his career. The two of them worked together in establishing the Modern Medical School of Vienna.[34] Škoda had a keen interest in Laennec's work, and in 1835 he began giving clinical courses on mediate auscultation in Vienna. His courses drew students from across the German-speaking world. After he published the first edition of his treatise on percussion and auscultation in 1839, both these techniques became better known and more widely used in Germany and beyond. But the historian Jens Lachmund emphasizes that 'Auscultation did not remain the same during its transportation from the Parisian clinics to Vienna and then generally to the German-speaking countries.'[35] It underwent changes, largely because of revisions to the technique made by Škoda. These changes were substantial enough that they produced a distinctive 'Viennese approach' to using the stethoscope.

What was the nature of Škoda's approach? How did it differ from Laennec's? Laennec proposed that there was often a direct correspondence between particular sounds and underlying pathologies. He considered many sounds to be 'pathognomic signs', that is, distinctive of particular pathological entities and existing in a one-to-one relationship with them. Although Škoda also recognized the link between sounds and anatomical changes, he was less convinced that pathognomic relationships existed between them. He argued instead that auscultation sounds should be explained by reference to the physical principles that produced them. He conducted

Joseph Škoda,
1850, lithograph
by E. Kaiser.

experiments on corpses and body parts to show how a variety of organic changes inside the body could create the conditions for similar acoustic effects to occur. This difference might seem subtle, but Škoda's approach had important implications and led to a simplified system for the classification and description of sounds as diagnostic signs.

Laennec's approach had placed great weight on the listener being able to detect subtle differences between sounds, and these subtleties translated into important differences in the diagnosis. Škoda believed that a taxonomy as sophisticated as that of Laennec was unwarranted, as not all sounds could be indexed to distinct anatomical changes. It was also impractical, as it required auditory judgements that were too fine for most people to make with a sufficient degree of reliability, so Škoda constructed his own simpler taxonomy of auscultation sounds.[36] For example, Laennec had identified three variants of amplification of the voice (heard through the chest if listening with the stethoscope while a patient was speaking). These included pectoriloquy, bronchophony and egophony, each being linked to particular pathologies. Škoda abolished these categories. His system of classification had only one sound, bronchophony, which could be either loud or soft. This sound was explained by reference to the physical principle of resonance of the voice in the lung, which could potentially be caused by a number of different pathological changes.

The importance which Laennec placed on an auscultator's capacity to detect subtle differences in auditory qualities and characteristics is reflected in the effort he makes in his treatise to construct careful and detailed descriptions of sounds. The fact that Škoda used a simpler taxonomy of sounds meant there was less need for the elaborate vocabulary and description, which many felt made Laennec's work difficult to read and hard to assimilate. When he did describe sounds, Škoda worked to use linguistic conventions that would be familiar to his German-speaking audience, making mediate auscultation more accessible and hence more readily available for use.

Škoda's work was influential. As had been the case in Edinburgh, he established Vienna as a centre for teaching and research in mediate auscultation that was independent of Paris. Although it was almost entirely overlooked in France, Škoda's taxonomy of auscultation sounds was integrated into German textbooks and ultimately became 'part and parcel of the German physician's diagnostic competence'.[37]

Lachmund suggests that a consideration of the Viennese approach to mediate auscultation gives us a valuable insight into the standardization of the use of the stethoscope. He identifies a tension between universalization and localization. In broad terms one might think of stethoscopic listening becoming a universal practice as it began to be taken up by doctors across Europe and even further afield. But while there may have been general similarities in the use of the stethoscope, local variations or 'vernacular forms' also appeared, shaped by the cultural, linguistic and material resources available in each particular medical context.[38] The Viennese approach was one such vernacular. This variation does not necessarily represent a failure of standardization. While techniques and technologies need to have broad applicability in order for standardization to become possible, they also need to be sufficiently versatile that they can be made relevant in a wide range of situations. Mediate auscultation had the combination of generalizable applicability and versatility that allowed it to be adopted internationally in broadly similar ways, while being moulded into locally appropriate forms, as we see in the cases of its use outside Europe, for instance in India and the United States.

Bengal

The stethoscope was introduced into Ayurvedic medicine in the late nineteenth century in Bengal. The historian and sociologist of science Projit Bihari Mukharji charts the development from early accounts of the stethoscope in textbooks from 1894 towards general acceptance of the instrument. During this time there was an expanded sensory understanding of diagnosis which advocated for examination by 'listening' (or *srotra pariksha* in Sanskrit), not only to patients' stories but through practices using a *nadiyantra*, or stethoscope.[39]

There were complementary practices in Ayurvedic medicine at the time of the stethoscope's introduction – such as pulse examination, which entailed close examination of bodily signs – and there was an understanding of the localization of disease within the body that fitted with its use. Those who did object largely did so in regard to the perceived futility of trying to achieve diagnostic precision, rather than the reliability of the stethoscope itself. What was more important, the objectors thought, was curative value.

Nonetheless, a role for the stethoscope was seen in Ayurvedic medicine. One of the main reasons that its reception may have been smooth was that there were no diagnostic techniques or technologies (such as immediate auscultation) which it was directly replacing, as there had been in the case of Europe. There were few anxieties about loss of skill through instrumentation and few concerns about the quality of the sound heard through the stethoscope compared to other ways of listening. The main obstacle to the tool's introduction was a need to transform the stethoscope's identity into one which was more consistent with Ayurveda and distinct from biomedicine.[40]

As the Kavirajes (as physicians are known in the Bengal region) adapted the stethoscope to their needs and situation, the stethoscope also changed Ayurvedic understandings of the body. In some texts, for example, it is possible to see how organs such as the lungs became more conspicuous through descriptions of clinical examination with the stethoscope, acquiring a more concrete and discrete reality than they had had previously.[41] The Ayurvedic lung is a different one, however, to that which the biomedical doctors were listening to. Mukharji explains:

> What is heard through the stethoscope then is a sonic lung, but one whose soundings are different from the sonic lungs that biomedical doctors heard through their stethoscopes. The difference lay in the fact that the Kavirajes, when they heard the sonic lungs, heard in them the auricular traces of *doshes* [a Sanskrit word for that which causes illness], whereas biomedical doctors perceived merely the movement of air and fluids.[42]

Just as Chinese physicians were sensing different qualities of a pulse compared to Greek physicians, because of their fundamental understandings of what a body is and how it works,[43] so too did the Bengali physicians listen to different lung sounds from European ones. In Bengal the stethoscope brought into view a distinctive technique and a brand new 'quasi-biomedical' pair of lungs. What happened in Bengal with the introduction and incorporation of the stethoscope, therefore, is an integration of a multiplicity of health beliefs, therapeutic knowledge and bodily practices.[44]

The United States

The adoption of the stethoscope in the United States is acknowledged to have been slow, and it took the entirety of the nineteenth century for the technique to become widespread.[45] The historian Richard Reinhart has explored the reasons for such a gradual uptake. He points out that some American physicians did attend lectures and clinics in Paris hospitals. They carried the technique of mediate auscultation back to their respective medical schools and practices. These physicians, however, were representatives of an academic medical elite, and training in auscultation subsequently became integrated into the curricula of the medical schools in the northeastern United States (such as Boston, Philadelphia and New York) with which they were associated. Physicians who were educated in these schools were likely to have been introduced to the stethoscope, and to have gone on to practise mediate auscultation. Most American physicians in the nineteenth century, however, received their training at so-called 'proprietary' medical schools which were owned and run by one or more doctors and were not affiliated to universities. These physicians, who were the majority, lived and practised in rural settings outside major urban centres of the American northeast. They would have been unlikely to encounter the stethoscope during their training, and afterwards had limited opportunities (and perhaps little inclination) to travel and learn new techniques. Some perhaps saw no great value in acquiring stethoscopic knowledge since it had little direct bearing on treatment options for their patients. A particular social geography, then, as well as structural factors underpinning the American medical education system, produced unevenness in the distribution of stethoscopic knowledge in nineteenth-century America.

Reinhart suggests that medical journals offered one means by which physicians could continue to learn and develop new skills after or outside of medical school. Comparing medical journals from the urban/northeastern and rural/western/southern United States, however, he notes that in the early and mid-nineteenth century the journals published in Boston and Philadelphia had more articles containing the words 'auscultation' and 'stethoscope' than their counterparts published in Louisiana and Kentucky, though later in the century the difference was only slight. While it is hard to know whether or not the readership of these journals was chiefly drawn from the same region as their publication, Reinhart found that the authors of the material for the northeastern journals were predominantly northeastern physicians (who were likely to have had well-developed knowledge of the stethoscope), while the authors of the journals published in Louisiana and Kentucky mainly consisted of regional physicians who were less likely to have had such knowledge.[46] The content of the journals, then, reflects the uneven distribution of stethoscopic knowledge and the uneven opportunities for stethoscopic learning outlined above.

In the early nineteenth century, medical education in America was unstandardized. Reinhart writes that it could range 'from as little as several months of didactic lectures repeated in the second year, supplemented by an apprenticeship with a practicing physician, to a formal training leading to an MD degree from a university-affiliated medical school'.[47] In 1850, for example, there were 201 medical practitioners in eastern Tennessee, of whom only 17 per cent were medical school graduates.[48] It was not until the end of the nineteenth century that American medical school curricula became more standardized. Also at that time, American hospitals underwent transition, shifting from being solely charitable organizations for the administration of care to being sites for scientific medical research. This scientific orientation meant the clinicians were increasingly using medical technology, including the stethoscope, in their work, and were honing their skill with these instruments. The techniques and skills used in the hospitals also found application outside them, so hospitals played an important role in the incubation and distribution of stethoscopic knowledge. As proprietary medical schools were closed and replaced by university-affiliated ones, hospitals were also increasingly used as sites of medical education, allowing medical students access to the bedside training they needed to develop as stethoscopic listeners. Despite these changes, however, in the nineteenth century the majority of Americans still lived in rural locations and their physicians had been locally trained. Many had not had the opportunity to gain experience in stethoscopic listening. As late as 1903, William Osler (1849–1919), a distinguished Canadian physician who among many other accomplishments had helped to establish the Johns Hopkins School of Medicine, remarked in a speech to the New Haven Medical Society: 'It seems shocking to say, but you all know it to be a fact that many, very many men in large practice never use a stethoscope.'[49]

Modification

We have outlined how the spread of mediate auscultation created variations in where and how the technique was practised. It also produced variations in the design of stethoscopes themselves. In the European context, stethoscopes could be bought in London as early as 1819. They were initially imported from Paris, but when supplies had run out, London woodturners such as Allnutt of Piccadilly and Grumbridge of Poland Street began to make and sell them.[50] By the time Laennec published the second edition of his treatise in 1826, his stethoscopic design had been simplified by removing the screw joint which had allowed the instrument to be separated into halves.[51] Although this modification reduced the portability and versatility of the stethoscope, it made the instrument cheaper and easier to produce.

In 1828, a colleague of Laennec's named Pierre-Adolphe Piorry (1794–1879), who's primary interest was in percussion but who also used auscultation, reduced the instrument's stem to 'the size of a finger'.[52] Later on, one type of the smaller so-called 'Piorry' stethoscopes was designed so that it could be clipped to the inside of a top hat. Setting the scene in one of the short stories in his collection

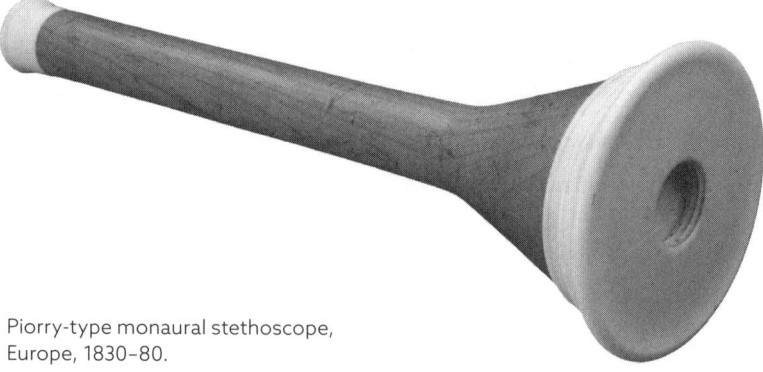

Piorry-type monaural stethoscope,
Europe, 1830–80.

Pierre-Adolphe
Piorry, 19th century,
lithograph by
N. E. Maurin.

Round the Red Lamp, published in 1894, Arthur Conan Doyle refers
to this kind of stethoscope:

> It is after one of the quarterly dinners of the Midland Branch
> of the British Medical Association. Twenty coffee cups, a
> dozen liqueur glasses, and a solid bank of blue smoke which
> swirls slowly along the high, gilded ceiling gives a hint of a
> successful gathering. But the members have shredded off
> to their homes. The line of heavy, bulge-pocketed overcoats
> and of stethoscope-bearing top hats is gone from the hotel
> corridor.[53]

Also, in 'A Scandal in Bohemia' from the *Adventures of Sherlock
Holmes*, Holmes is able to deduce that Dr Watson, whom he has not
seen since Watson got married, has resumed his medical practice:

if a gentleman walks into my rooms smelling of iodoform, with a black mark of nitrate of silver upon his right forefinger, and a bulge on the right side of his top-hat to show where he has secreted his stethoscope, I must be dull, indeed, if I do not pronounce him to be an active member of the medical profession.[54]

In 1879 an Edinburgh medical student was accused of possessing a dangerous weapon when his stethoscope fell from his top hat during a snowball fight.[55]

Another variation on Laennec's design came in 1828, when Nicholas Comins (dates unknown), a physician to the Edinburgh Royal Infirmary, produced a flexible monaural stethoscope. Comins was concerned by the discomfort auscultating with a wooden cylinder caused to patients, as they were required to make a variety of changes of position during an examination and also had to accept the pressure of the instrument against their bodies. These drawbacks were a particular problem, Comins felt, in the teaching context, 'in consequence of the torture unavoidably inflicted by repeated attempts, and by the frequent change of posture necessarily required of the afflicted persons'.[56]

In order to get around these problems, Comins designed a stethoscope which consisted of two tubes connected by a flexible joint which would allow the stethoscopist 'to explore any part of the chest, in any position, and in any stage of disease, without pressure or inconvenience to the patient or to himself'.[57] Nowadays, even when students are armed with modern flexible stethoscopes with rubber tubing, patients sometimes object to large groups of them listening to their chests on account of the invasiveness of the procedure and the need for them to expend precious energy by changing position to make listening possible. Flexible stethoscopes, then, did not

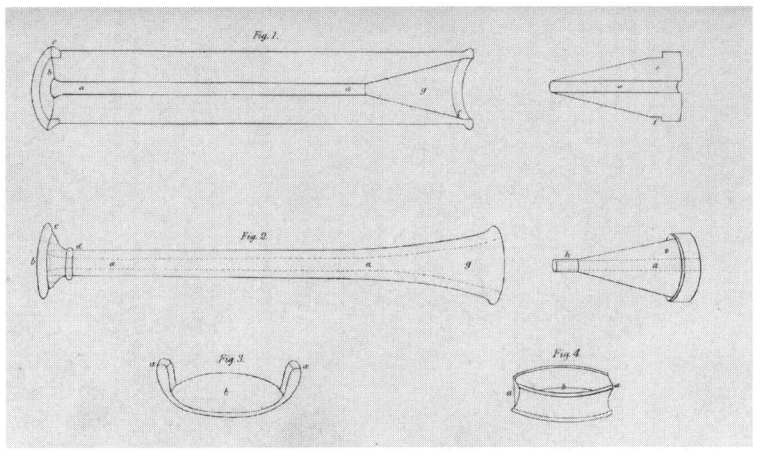

Construction of the stethoscope, illustrations from Charles J. B. Williams, *The Pathology and Diagnosis of Diseases of the Chest*, 3rd edn (1835).

entirely eliminate the drawbacks Comins originally associated with non-flexible ones.

At around the same time as Comins, others were also putting forward designs of flexible stethoscopes involving tubes of coiled brass or iron wire covered with cloth, and in 1829 one of Laennec's English pupils, Charles James Blasius Williams, attempted to create a binaural stethoscope: 'It had a trumpet-shaped chest piece of mahogany, the end of which screwed into a connection to which were added two bent lead pipes.'[58] The lead pipes could be adjusted to fit the ears, but there were no earpieces and the device was difficult to use. In 1851 Dr Arthur Leared (1822–1879) produced a binaural stethoscope which had a rigid chest piece connected to two flexible pipes made of gutta-percha, a type of hard, tough rubber, reaching to the auscultator's ears. Other so-called 'improved versions' of the binaural stethoscope were also produced that year, but were not very practical, requiring three hands to adjust and use.[59]

In 1855, thanks to improvements in the manufacture of rubber, Dr George Philip Cammann (1804–1863) was able to publish a design

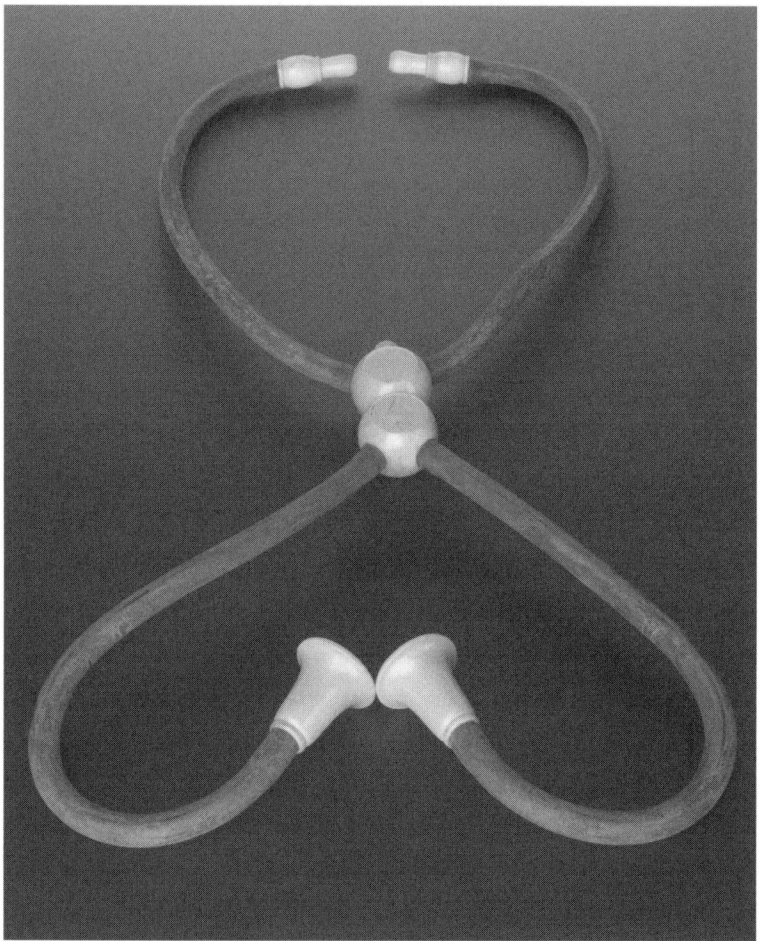

Binaural stethoscope, England, 1910–40.

that used flexible tubing to conduct the sound from a single chest piece into two earpieces. This stethoscope was light and could be carried easily, and the rubber tubes meant the instrument could be folded into a pocket. It became a standard design and was mass-produced. Modification and innovation continued (as we will discuss in later chapters), but by 1855 the stethoscope was in more or less its modern form. Monaural instruments, however, were in use long after Cammann's design became available, and many doctors preferred them. Metal monaural stethoscopes were being used by some German army officers in the Second World War, and Russian-made versions were captured during the Korean War, which took place between 1950 and 1953.[60]

The binaural stethoscope with flexible tubes seems at once so near and so far from Laennec's original paper cylinder. The fundamentals of the device and its application have remained the same, but different social and cultural contexts as well as new materials and individual innovations have created a wide variety of forms. 'The rise of the stethoscope', then, might be better understood as 'the rise of the stethoscopes'. It is important to remember, however, that modifications were sometimes dismissed. In Bengal, for example, updates of the stethoscope were sometimes seen as luxurious and superfluous. Authors of textbooks on the stethoscope often sidestepped the issue of the merits of different versions of the instrument so that their advice remained relevant to those, often working among the rural poor, who could not access the latest models. One author criticized biomedical doctors for having turned the instrument into an ornament.[61]

Despite pockets of resistance and unevenness in uptake, and variations in the use of modified designs, in general more and more doctors acquired and used stethoscopes. During this time acoustic understanding of the body became wider, deeper and more precise. But the stethoscope did not just have transformative effects upon the body as an object of medical knowledge. It also had profound implications for the lived bodies of patients and doctors. Both had to adjust in their own ways to a new technology, and the stethoscope became a powerful mediator of the relationship between them.

3
Reach

As the stethoscope started to spread around the world it also spread around the body. New areas and their sounds were charted and mapped. Laennec's key expertise had lain in diseases of the chest, and his research focused on the application of the stethoscope to this area. Colleagues of Laennec's who had different interests and specialisms, however, soon began to apply the instrument to other parts of the body and to make it indispensable to their fields of expertise. As use of the stethoscope became more widespread and intensive, medical understanding of the variety of sounds that the body could produce became more detailed and comprehensive. Stethoscopists revealed a vast repertoire of sounds that had immediate or potential diagnostic value.

Although doctors had previously listened to foetuses using immediate auscultation, Laennec's contemporary and friend Jacques-Alexandre Le Jumeau de Kergaradec began experimenting with mediate auscultation on women in pregnancy. He writes:

> One day whilst examining a patient near term and trying to follow the movements of the foetus with the stethoscope I was suddenly aware of a sound that I had not noticed before; it was like the ticking of a watch. At first I thought

I was mistaken, but I was able to repeat the observation over and over again. On counting the beats I found that these occurred 143–148 times per minute and the patient's pulse was only 72 per minute.[1]

After making this observation in numerous other cases he concluded that the sound he could hear was the foetal heartbeat. He also heard a blowing sound that was synchronized with the beating of the mother's heart. This sound became known as the 'placental souffle' and is made by the flow of blood through the placenta. Kergaradec's findings had considerable implications for obstetrics. Most obviously, he had identified sonic signs which facilitated the diagnosis of pregnancy. They could also be used to determine whether a foetus was alive even when it was not moving, and to ascertain the presentation and position of the foetus both for vaginal deliveries and Caesarean sections. The presence of two sets of foetal heart sounds indicated twins, and if the sounds were heard in areas where they were not usually expected, they could point to ectopic pregnancy, where the foetus develops outside the uterus, often in a fallopian tube.

Laennec praised Kergaradec's work and cited it at length in the second edition of his treatise. It was also generally well received in academic circles, though one elderly obstetrician wrote an open letter in which he dismissed the stethoscope as a 'newfangled and ridiculous plaything' and recommended that Kergaradec 'abandon these toys of ignorance truly prejudicial to science and to the well-being of an amiable and interesting sex'.[2] As had been the case with mediate auscultation more generally, the acceptance of the stethoscope by male midwives or accoucheurs (midwifery at this time was a male-dominated occupation) was not always straightforward. Some refused to believe that it was even possible to hear the foetal heartbeat, and others felt that the use of mediate auscultation on pregnant women went beyond the bounds of propriety. The value of the stethoscope was gradually accepted, however. For instance, after studying first in Edinburgh and then in Paris under Laennec and Kergaradec, Irishman John Creery Ferguson (1802–1865) promoted the use of the stethoscope upon his return to Dublin. By 1830 the instrument was in daily use at the Rotunda Hospital, a lying-in hospital and a centre for education in midwifery in the city.[3]

In 1834, a German obstetrician named Anton Friedrich Hohl (1789–1862) designed a special stethoscope for listening to the foetus. Horn-shaped and shorter than Laennec's cylindrical design, it was thought to transmit the foetal heart sounds more clearly. A reworking of Hohl's design was put forward by Jean Depaul (1811–1883) in 1847, and in 1895, Adolphe Pinard (1844–1934), another French

Adolphe Pinard, 19th century, photograph by Pierre Petit.

Pinard-type foetal
stethoscope,
London, 1870–1920.

obstetrician, gave his name to the 'Pinard horn', essentially a cone of wood, metal or plastic, about 20 centimetres (8 in.) long. Because it had to be pressed close to the foetal heart in order for the heartbeat to be heard, it gave a very accurate indication of foetal position. This instrument is still sometimes used by midwives today. There has even been a resurgence of interest in auscultation as a primary method of foetal surveillance during labour due to anxieties over the excessive medicalization of childbirth and the overuse of technology.[4] Some parents also find listening to their unborn baby's heartbeat during pregnancy to be reassuring and to help with bonding, and some websites recommend the use of conventional stethoscopes for this purpose.

While Kergaradec had listened to the pregnant belly, another colleague of Laennec's, Jacques Lisfranc (1790–1847), applied the stethoscope to the bones. He noticed that he could hear so-called 'crepitations' (creaking, cracking, grating or crunching sounds) at the site of fractures much more clearly than had been possible with the naked ear. Summarizing Lisfranc's key findings, a review in the *American Journal of Medical Sciences* explains some of the sonic qualities of different types of fracture:

> The crepitation produced by fragments of the compact bones, furnishes harsh, rough sounds, strong cracklings, perceived by the stethoscope, they are often piercing, and sometimes greatly fatigue the ear . . . The crepitation of fragments of spongy bones is dull, and similar to the action of a file upon a hard and porous body, (pumice stone for example;) . . . The crepitation of oblique fractures is stronger than that of transverse fractures . . . If liquids be poured out around the fragments, a noise similar to that which the foot produces in an old shoe that contains water, is added to the crepitation . . . When the fracture is complicated by splinters, there is heard, with the ordinary crepitation, a sort of crackling, similar to that which a number of hard and angular bodies would furnish, if rubbed the one against the other.[5]

The stethoscope, Lisfranc argued, could identify the exact position of fractures in bones all over the body, including the tibia, fibula, patella, femur, pelvis, radius and ulna, humerus, clavicle, scapula, vertebrae, jaw and skull.

As with Kergaradec, Laennec referred to Lisfranc's work at some length in the second edition of his treatise. Although the auscultation of fractured bones does not appear to have been adopted extensively by later doctors, several sources suggest that some did find it valuable, and recent commentaries indicate that the technique is still occasionally used today. The stethoscope can also be applied to the joints, which as we are all well aware can sometimes be heard creaking, clicking and cracking. An article written in 1911 claims that auscultation 'sometimes reveals the commencement of disease in joints which to inspection and palpation appear to be normal, and which may even cause no discomfort to their possessors'.[6] It indicates the presence of organic change before any symptoms manifest themselves.

In addition to his work on bones, Lisfranc applied the stethoscope to the pubis or sacrum in the diagnosis of bladder stones. He argued that the friction sound made when a catheter inserted into the bladder made contact with a stone could be heard more distinctly using a stethoscope than was otherwise possible. This finding was endorsed by Laennec:

> When the stethoscope is applied to the os pubis or sacrum while the catheter is introduced, we hear the sound occasioned by this coming in contact with the stone, much more distinctly and loudly than we can do with the naked ear; and, indeed, even in the obscurest cases, the sensation communicated will be quite as distinct as would be that produced in the open air by striking the instrument, even much more forcibly, against a stone.[7]

Joseph Škoda, who founded the Viennese approach to auscultation, also confirmed this finding. He mentioned the application of the stethoscope to the pubis in the diagnosis of bladder stones in his treatise on percussion and auscultation, arguing that the stethoscope supported a diagnosis which would otherwise depend upon tactile sensations produced using the catheter: 'Auscultation must be considered of some value in the detection of urinary calculi. It aids and supports the sense of touch.'[8]

Other parts of the body were also subjected to stethoscopic examination. Laennec described pressing the stethoscope to the mastoid process, a bony projection behind the ear. He asked patients with hearing difficulties to breathe in sharply through the nostril on the same side while holding the other nostril closed. If there was any moisture in the Eustachian tube a 'guggling' sound could be heard, while if the tube was blocked there would be no sound at all.[9] These acoustic signs gave indications as to whether or not it would be appropriate to try to attempt to cure deafness by clearing the blockage. The stethoscope was also applied to the head more generally. In 1833, a Boston physician named John D. Fisher read a paper on the subject of the 'cephalic bellows-sound', which he had discovered in certain diseases of the brain.[10] Doctors have continued to employ cranial auscultation. Writing in 1955, London neurologist Ian Mackenzie describes how this examination should be conducted:

A bruit should be listened for, in quiet surroundings, over the skull and eyeballs, the latter situation being the most favourable for hearing the softest ones. The patient should be asked to close both eyes gently and the stethoscope firmly applied over one eye. During auscultation the other eye should be opened as in this way there is considerable diminution of eyelid flutter, which may cause confusion if rhythmical. Auscultation is then carried out over the other eye in a similar manner. If a murmur is not readily heard the patient should be asked to hold his breath. Finally,

auscultation should be carried out over the temporal fossae
and mastoid processes.[11]

Orbital bruits, sounds caused by occlusions or blockages to the arterial blood supply in the brain, are still considered valuable diagnostic signs.

Among other sounds heard through the stethoscope were borborygmi, an onomatopoeic term for gurgling noises in the gut. These are still useful as an indicator that the digestive processes are active, and that matter is moving through the intestines. These particular sounds had, of course, been familiar for centuries, but became included in an ever-widening repertoire of auscultatory sounds. The eventful soundscape inside the body was characterized by flow and movement of blood around the body, of breath in and out of the lungs, and of matter and gases through the gut. There were also sounds of contact and friction between tissues and bones, and of perfusions of gas through liquids. While many sounds were produced involuntarily by the body, others required the active involvement of patients (who might, for instance, be required to speak, take deep breaths or hold their breath, inhale sharply through one nostril and so on while the doctor listened). Still others required the skilled manipulation of patients and instruments by the doctor, as in the case of listening for the sound of the catheter against a bladder stone. But it is important to note that while a living body might be noisy, a silent body equates to death. The medical students with whom Tom conducted his research were taught that one of the steps that should be taken in the formal confirmation of death is careful listening for heart and breath sounds. These would, of course, be absent if a patient had died.

For some, the fact that the stethoscope can be used to detect signs of the very beginning and also the end of life lends the instrument particular power. In his poem 'The Stethoscope', the Welsh poet and physician Dannie Abse writes:

> Through it,
> over young women's tense abdomens,
> I have heard the sound of creation
> and, in a dead man's chest, the silence before creation began.

There is a sense in Abse's poem that the stethoscope reaches, or allows him to reach into, a spiritual or cosmic space.[12] At the same time, because of its close association with cardiological medicine, the stethoscope is often construed as a device for listening to 'hearts', generally considered to 'lie in the centre' and, in Western thought, to represent the inner seat of feeling and emotion. Once during his fieldwork a medical student told Tom he would be taking his 'emotional stethoscope' with him when he went to see a particular friend of his that evening, because he couldn't tell whether or not she had feelings for him and wanted to know. He explained that the stethoscope was, after all, 'a device for listening to hearts'. It is an instrument for obtaining depth, enabling the user to get past an outer layer and into the core.

Adjusting to the Stethoscope

Listening with a stethoscope created interpersonal challenges for doctors and patients. For instance, while less intrusive than immediate auscultation, mediate auscultation nonetheless was, and indeed still is, intrusive to an extent. It involves the doctor touching the patient, even if only indirectly, via the stethoscope. It sometimes requires the patient to be partially undressed in the presence of the doctor, or at least to hold aside a piece of clothing so that the

Kaisu Koski, *Listening Gaze*, 2008, installation exhibited at De Veemvloer, Amsterdam.

stethoscope can be pressed to the exposed skin or to a thin layer of underclothing. In England, this was a threat to Victorian notions of decency and decorum. It has been suggested that Queen Victoria herself had a great aversion to the stethoscope (she apparently resisted undergoing physical examination in general and her own doctor was obliged to treat her following only verbal communication).[13] The historian of medicine Stanley Reiser mentions a stethoscope which was designed to preserve physical and social distance between doctor and patient. It had a tube several feet long to allow the patient to hold one end onto their body while the doctor listened in another room.[14] Patients also often had to change position and breathe, cough or speak at the instruction of the doctor. This was very different from medical interactions based around a verbal questioning, and took some getting used to: 'Patients had to learn to comply with these procedures technically and they had to accept them as moral and beneficial.'[15] Doctors in turn had to develop an etiquette through which to introduce the stethoscope and explain its purpose to their patients.

Those who were among the first patients to encounter the stethoscope were understandably disconcerted by it. The medical historian P. J. Bishop describes the stethoscope as 'the first major diagnostic tool medicine ever had'.[16] Medical instruments were associated with surgeons, not physicians, so when the doctor pulled out a stethoscope during a consultation some patients feared a painful procedure might be imminent.[17] There were other reasons, too, for patients to be apprehensive of the stethoscope. After all, a great deal can hinge on what a doctor hears with it. Patients could not control what their bodies would reveal. They could not know what the listening doctor was hearing or the implications of his findings. This feeling of exposure and vulnerability in the face of medical judgement is one to which we can all relate. An article in *The Times* published in 1824 refers to the anxieties the use of mediate auscultation might produce for the patient:

> It is quite a fashion if a person complains of a cough, to have recourse to the miraculous tube, which, however, cannot effect a cure; but should you unfortunately perceive in the countenance of the Doctor, that he fancies certain symptoms exist, it is very likely that a nervous person might become seriously indisposed and convert the supposition into reality.[18]

The fact that there was originally a strong association between the stethoscope and the diagnosis of tuberculosis probably did not help to dispel fear of the instrument, and perhaps due to its general association with illness and the seriousness of medical judgement, fear of the stethoscope never entirely subsided. 'To the Stethoscope', for instance, is a heavily melodramatic five-page poem published in 1847:[19] 'As an instrument on which the hopes and fears, and one may also say the destinies of mankind, so largely hang', writes the poet (identified only as V. V.), the stethoscope 'appears to represent a fit object for poetic treatment'.[20] The opening lines of their own effort read:

> Stethoscope! Thou simple tube,
> Clarion of the yawning tomb,
> Unto me thou seem'st to be
> A very trump of doom.
>
> Wielding thee, the grave physician
> By the trembling patient stands,
> Like some deftly skilled musician;

Strange! The trumpet in his hands.
Whilst the sufferer's eyeball glistens
Full of hope and fear,
Quietly he bends and listens
With his quick, accustomed ear –
Waiteth until thou shalt tell
Tidings of the war within:
In the battle and the strife,
Is it death or is it life,
That the fought-for prize shall win?

The stethoscope is positioned as a kind of messenger, indicating life or death (though it quickly becomes apparent that the poet is heavily preoccupied with the latter). The instrument seems to speak of its own accord, communicating secretly with the doctor.

The poet goes on to imagine a 'gentle maiden' lying sick on a couch. She does not recognize the seriousness of her symptoms, and thinks her illness 'but a passing'. The stethoscope, however, thinks and seems to speak differently:

Then, thou fearful stethoscope!
Thou dost seem they lips to ope,
Saying, 'Bid farewell to hope:
I foretell thee days of gloom –
Make thee ready for the tomb!'

The poem is shot through with morbid and macabre imagery from religion and mythology. The figure of Death arrives to claim this patient in grim wedlock.

The poem goes on to describe a 'pale youth', lying awake, dreaming of the future and all that he will accomplish in it, imagining huge achievements in fields such as poetry, painting, sculpture, music, politics and the natural sciences. But:

> thou doleful Stethoscope!
> Thou dost seem to say,
> 'Hope thou on against all hope,
> Dream thy life away:
> Little now there is to spend;
> And that little's near an end.
> Saddest sign of thy condition
> Is thy bounding, wild ambition;
> Only dying eyes can gaze on so bright a vision.'

There then follows a list of a variety of symptoms (which seem consistent with tuberculosis, though no disease is named), which will bring the pale youth to the verge of death. Satan is imagined to appear and lurk around the young man, encouraging him to doubt the existence of God and the afterlife. Then God, Christ and 'the blessed Spirit' are invoked to give the patient aid and support. We never learn of the outcome. The poem concludes with an image of Life and Death personified, wrestling over who will have, and play upon, the stethoscope. Ultimately the two strike a truce: 'Life shall sometimes sound a blast' but 'King Death' will 'blow/ With no bated breath./ Long drawn out, and deep and slow/ Shall the wailing music go.' The stethoscope can occasionally offer hope, but for the poet it is essentially a memento mori.

We also see evidence of the stethoscope striking fear into the hearts of those who behold it in Gustavus Hindman Miller's *Dictionary of Dreams*, published in 1909. The 'Stethoscope' entry is dark: 'To dream of a stethoscope, foretells calamity to your hopes and enterprises. There will be troubles and recriminations in love.'[21]

Stethoscopic examination and auscultatory findings also mark a sinister turning point in the narrative of Thomas Mann's novel *The Magic Mountain*. Published in 1924 but set in the years prior to the First World War, the narrative follows a young man in his early twenties named Hans Castorp. He goes to visit his cousin Joachim Ziemssen, who is suffering from tuberculosis and is being treated at a sanatorium in Davos. Castorp's visit is supposed to be brief, but he is repeatedly prevented from leaving to return to his life in the 'flatlands' by ill-health. Eventually, Castorp decides to accompany his cousin to a routine examination and to consult with the sanatorium's chief doctor, Hofrat Behrens. When they arrive, the doctor is waiting expectantly, 'standing in the middle of the room, in his white smock, holding the black stethoscope in his hand and tapping his thigh with it'.[22] He begins by examining Joachim but afterwards turns his attention to Castorp:

> He tapped all over, as he had done with Joachim, and several times went back and tapped again … 'Breathe deep! Cough!' commanded the Hofrat, who had taken up the stethoscope again; and Hans Castorp worked hard for eight or ten minutes, while the Hofrat listened. He uttered no word, simply set the instrument here or there and listened with particular care at the places he had tapped so long. Then he stuck the stethoscope under his arm, put his hands on his back, and looked at the floor between himself and Hans Castorp.[23]

The examination and auscultatory findings bring about Castorp's heavily foreshadowed transition from visitor to patient at the sanatorium.

Despite provoking embarrassment and anxiety among some, many patients who encountered the stethoscope early on were impressed and even amazed by the seemingly magical powers the instrument seemed to give the physician, granting them insight into what was happening inside the body (though some doctors worried that listening to patients' bodies through a wooden cylinder, as if the disease were a living entity that could speak, might make them appear foolish).[24] By the second half of the nineteenth century, patients in general were getting more accustomed to the stethoscope. They started expecting it to be used on them as part of a thorough examination. A number of doctors to whom we spoke during our fieldwork said they felt this remains true today. Respiratory and cardiac patients in particular expect to be listened to, and question their doctors if for some reason the stethoscope is not used during a consultation.

Stethoscopy's Golden Age

The nineteenth century has been referred to as 'the golden age of stethoscopy'.[25] The technique became increasingly accepted and widely practised as the century wore on. Physicians were not only applying the stethoscope to more areas of the body but were identifying new sounds. They were also linking heard sounds with increasing confidence to underlying organic changes and physiological events. For instance, in relation to the heart, the Irish physician Dominic Corrigan (1802–1880) gave a full description of the double murmurs sometimes heard in patients with a diseased aortic valve. He described how the first bellows murmur or *bruit de soufflet* was produced by blood rushing through the diseased valve up the aorta, and the second by blood rushing back through the valve into the ventricle.[26] The influential American physician Austin Flint (1812–1886) gave his name to a murmur occurring just before the first sound of the heart (the 'lubb' of the 'lubb-dupp'), which he described as being produced when blood is forced by the contractions of the auricles through a stiff or roughened mitral valve.[27] Clinicians also began to suspect that some murmurs and clicks might be related to problems with the heart's function, but might not have a clear organic cause.

Nineteenth-century clinicians were aided in their investigations of heart sounds by developments in physiology, that is, the study of functions and mechanisms in living systems such as the human body. Étienne-Jules Marey (1830–1904), for instance, made advances in techniques for studying and recording blood pressure, most notably in the design of the sphygmomanometer.[28] Although the German physiologist Karl von Vierordt (1818–1884) had produced versions of this instrument previously, in 1860 Marey introduced a more reliable and expedient design, which gave a line recording of the strength and rate of a person's pulse. Marey also used catheterization (the technique of inserting a cannula, or very narrow tube, into a blood vessel) to engage with the moving heart. Working mainly on animals, and particularly horses, he studied changes in blood pressure in the different chambers of the heart during the cardiac cycle. This kind of pioneering work helped make doctors more assured in their understanding of how sounds heard through the stethoscope related to the heart's action.

It is interesting to note that during the nineteenth century in England, auscultation was just part of a wider societal interest in sound and listening. There was the literary heritage of the Romantic poets, who had a 'preoccupation with the sublime force of the music and quiet of nature'.[29] There was considerable scientific experimentation on sound, and investigations were made into otology and the workings and afflictions of the ear. Important sound technologies were invented and introduced to society more widely, including the electric telegraph, the microphone, telephone and phonograph. These technologies all required new orientations to sound and specific 'audile techniques' or 'practices of listening that were articulated to science, reason, and instrumentality and that encouraged the coding and rationalization of what was heard'.[30] The Victorian era was, as the literary historian John Picker writes, 'an auscultative age'.[31] At the same time, we can understand the stethoscope as being at the origin of a genealogy of technologies that made listening directional and directed, and which constructed a private, individualized acoustic space for the listener. We see this genealogy continuing today in the form of ever new types of headphones and earbuds. They are iterations of earlier devices which have enabled listeners to be isolated in a sonic space and to focus closely on the characteristics of particular sounds; as such, they are expressions of a stethoscopic principle.

Making the Doctor

The development of physical diagnosis (the use of percussion and mediate auscultation) has been linked to the professionalization of medicine. Indeed, the medical historian Jens Lachmund goes as far as to argue that they were co-produced or co-constructed by one another. How could this be so?

Generally speaking, at the beginning of the nineteenth century the provision of medical expertise across Europe and America was something of a free-for-all. A wide variety of medical practitioners were active, 'untrained chemists, druggists and quacks', who competed directly with qualified physicians.[32] Those physicians who were academically trained were also in competition with one another. Medical expertise was largely seen as a quality of an individual doctor, and a doctor's success rested on his (at this time it was always a 'he') ability to gather and preserve a good reputation, convincing patients (and particularly wealthy and powerful patients) of his superiority over his rivals.

During the nineteenth century, however, doctors began to organize themselves and to build up professional strength, exerting political pressure for reform of this free market. In Britain, for example, doctors set up local medical societies, which called for 'bans on unqualified practice, and a single national register for all qualified medical men'.[33] They also called for the Royal College of Physicians, which issued licences to practise medicine, to be abolished or radically reorganized because they felt it worked to protect the status quo and the interests of elite metropolitan doctors rather than furthering the interests of physicians more generally. The Provincial Medical and Surgical Association, established in 1832, became the British Medical Association in 1857, and grew in membership, strength and influence. Meanwhile, medical licensing was undergoing gradual reorganization. The 1815 Apothecaries Act set a basic standard requirement of education for doctors in England

Swain, after a painting by William Small, *The Good Samaritan*, 1898-9, halftone of a physician examining a child with a stethoscope outside a tent on the side of the road.

Leather doctor's bag with contents, England, 1890–1930. The bag and its contents originally belonged to Professor John Hill Abram, the latter comprising a variety of medical instruments including stethoscopes and syringes.

and Wales. This piece of legislation was followed by the Medical Act of 1858, which created new qualification requirements for doctors, as well as a register for approved practitioners. It also instigated the creation of the General Medical Council, which was responsible for identifying and taking action against various kinds of medical malpractice. Although it was still not illegal to practise medicine without a licence, the existence of the register, and the threat of possible removal from it, served to push irregular practitioners to the medical margins and strengthened a sense of professional identity among trained doctors.[34]

In the German-speaking medical world, physical diagnosis (that is, percussion and mediate auscultation) played an important part in the drive for professionalization in medicine.[35] As we have discussed previously, these techniques could only be properly acquired through careful instruction, so a doctor's ability to use them became emblematic of formal medical training. Also, physical diagnosis had

William Strang, *The Stethoscope*, 1896, etching on paper. The candle-light here seems to suggest acoustic illumination.

arisen from a commitment to the tradition of pathological anatomy and to the application of rational empirical methods to diagnosis. As a result, their use required and expressed a certain coherence of thought and practice. Physical diagnosis represented a modern, scientific medical approach with which trained physicians identified. Using a stethoscope to perform mediate auscultation therefore became a symbolic act around which members of the medical profession could gather and through which they could present themselves to their clientele: 'as a material procedure it was highly visible and identifiable for both the lay-public and professionals themselves.'[36]

Physical diagnosis, then, became important in the building of the medical profession, and the stethoscope became the symbol of that profession as it established itself. This is an important part of the answer to the question of how the stethoscope came to be 'the doctor's symbol of office'.[37] The ongoing iconic nature of the instrument is something we explore further in the next chapter. Mediate auscultation contributed to the creation of a new self-consciousness in doctors, an awareness of their potential to be recognized as skilled and knowledgeable professionals. It might be argued that through the invention of the stethoscope, a particular kind of doctor was also invented.

The stethoscope can also be seen as an extension of the doctor's senses. It makes the sense of hearing and of sight more penetrating and powerful. As auscultation became part of routine practice for doctors, they came to rely on the stethoscope in order to exercise the full extent of their diagnostic competence. Doctors did not only develop a close relationship with the stethoscope as a symbol. The instrument also became an integral part of the medically competent body.

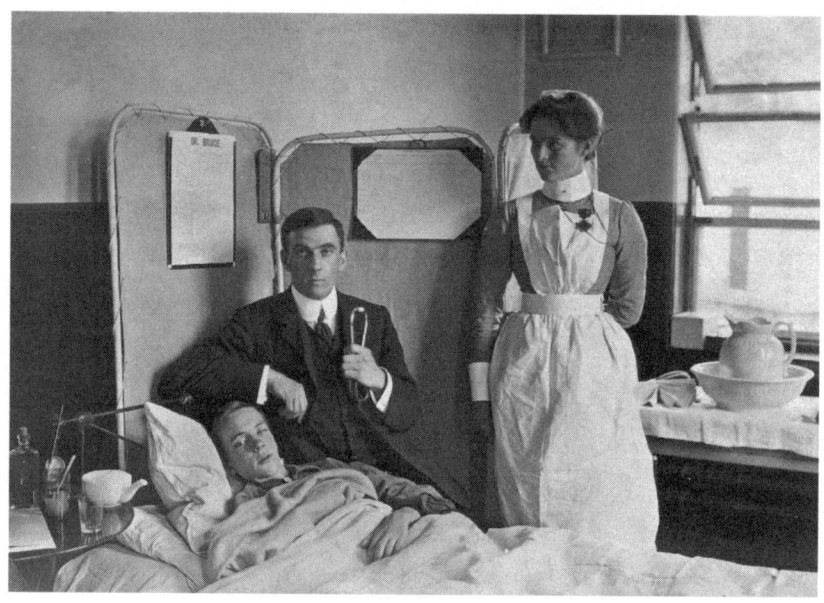

Basil Hood and nurse with a patient at London's Charing Cross Hospital, 1906.

4

Routine

During the twentieth century the stethoscope was being used to represent doctors in many public spaces, from advertisements, to book and magazine covers, to characters in movies and television shows.

A medical consultation was seen to combine both scientific and social authority, as is evident in a photograph of Dr Basil Hood with a patient and nurse at Charing Cross Hospital in London, taken in 1906. Dr Hood looks solemnly into the camera, his stethoscope held up to his chest. The gaze of the nurse seems deferential, as if she is attendant upon him, an extension of his agency rather than an actor in the medical situation in her own right. As we will see, however, nurses later came to wield stethoscopes themselves, and while the instrument remains firmly associated with medicine, it has been adopted by people in occupations and pursuits far beyond the medical sphere.

More than just a symbol and icon, however, during the early twentieth century the stethoscope was being used actively by doctors to diagnose diseases across the clinical spectrum. It continued to assist the diagnosis of, for instance, tuberculosis as well as conditions such as rheumatic heart disease, which occurs when rheumatic fever causes damage to the heart valves, often creating murmurs.

The stethoscope was also important for the detection of congenital heart defects.

We can find examples of the routine nature of stethoscopic listening in fiction from the mid-twentieth century. In 'Black as Egypt's Night', a semi-autobiographical short story, the Russian doctor, novelist and playwright Mikhail Bulgakov indicates that the stethoscope was indispensable to the physician at the time of its original publication between 1925 and 1927. Describing his waiting for a patient to come upstairs to his room, and not knowing what challenges the patient would present, he writes: 'My right hand lay ready on a stethoscope, as though on a revolver.'[1] Here the stethoscope seems to offer some reassurance. It is something to fall back on. But it is also something that he appears to feel he can use against the patient, or rather, against the difficulties the patient's case may bring.

Another analogy between the stethoscope and a weapon appears later in the story. The title 'Black as Egypt's Night' is a reference to the winter sky outside the remote country surgery where Bulgakov finds himself stationed. But it also alludes to the general lack of development in the area, and to what Bulgakov regards as the astonishing backwardness of the local people who come to the surgery. His colleagues tell him, for instance, about a patient of his predecessor's who was diagnosed with laryngitis and prescribed French mustard-plasters, one to be placed on his back between his shoulder-blades and the other on his chest. The man returned to the surgery after two days complaining that he felt no better. It emerged that he had placed the French mustard-plasters in the correct place, but on top of his clothes. Bulgakov is also told of the local practice of placing sugar crystals in the birth canal of women in difficult labour to entice the baby to come out. Later in the story Bulgakov encounters this local attitude for himself, when a man takes the ten doses of quinine he has been prescribed for malaria all at once rather than spaced out over ten days as prescribed, in the belief that this will shorten his treatment. Bulgakov is obliged to pump the man's stomach to get rid of the quinine, which keeps him up until dawn. He writes: 'After a hard night, sweet sleep overtook me. Darkness, black as Egypt's night, descended and in it I was standing alone, armed with something that might have been a sword or might have been a stethoscope. I was moving forward and fighting . . . somewhere at the back of beyond.'[2] Here the ambiguous sword/stethoscope in the dream is a weapon in a struggle against a vast darkness that includes disease and ignorance but also, it seems, Bulgakov's own sense of hopelessness, impotence and despair.

An Increasingly Powerful Profession

While now an established piece of a doctor's toolkit, during the twentieth century the stethoscope was increasingly being joined by other instruments. Doctors had a growing number of technologies at their disposal. There were thermometers, which could give measurements of body temperatures that could themselves be mapped on to fever charts to identify patterns associated with particular diseases. Sphygmomanometers allowed the measurement of blood pressure and the identification of abnormalities that could be linked to specific circulatory disorders. Laboratory tests and procedures began to enable those doctors with access to them to examine bodily fluids 'by measuring electrolytes, counting blood cells, searching for microbes, and observing unusual tissues and cell types'.[3] Use of these techniques supported doctors in the performance of their role.

Much as exhibiting sharp diagnostic skills and using emergent lab techniques may have helped to make a good impression on patients and allowed doctors to recognize the diseases from which their patients were suffering, in the early twentieth century there

were still few effective medicines and doctors knew that in most
cases they had little influence on the course of disease and on patient
outcomes. The mid-1930s, however, saw the introduction of effective
antibiotics in the form of sulfa drugs, which meant that in wealthy
countries major infections such as tuberculosis, meningitis and
rheumatic fever could now be treated. After the Second World War
the scope for effective treatment of infections improved still further
following the introduction of penicillin. Other new drugs were also
introduced which helped in the treatment of a variety of conditions
such as arthritis and high blood pressure.[4] The stethoscope, then,
was becoming symbolic of a profession with ever-growing therapeutic
power.

Research on auscultation continued with intensity throughout
much of the twentieth century. Increased specialism within the
medical profession meant that research focused ever more intensively
on particular organs, systems and diseases. In cardiology, auscultation
was, of course, concentrated on heart sounds. Doctors made closer
and closer investigations of murmurs, seeking to pin down their
acoustic characteristics, origin and place in the cardiac cycle, and
to link these sounds to normal and abnormal physiology and
anatomy. For instance, in 1933 the American doctor Samuel Levine
(1891–1966) developed a grading system from 1 (very soft) to 6 (very
loud) to describe the volume of systolic murmurs, which are turbulence in the flow of blood as it is pushed through the aortic and
pulmonary valves when the ventricles contract. This system is still
in use today. In the 1950s, the English doctor Aubrey Leatham
(1920–2012) demonstrated that the second heart sound (the 'dupp'
of the 'lubb-dupp') itself had two identifiable components, the first
produced by the closure of the aortic valve and the second by the
closure of the pulmonary valve. He was able to link an abnormal
splitting of these sounds to heart problems such as bundle branch

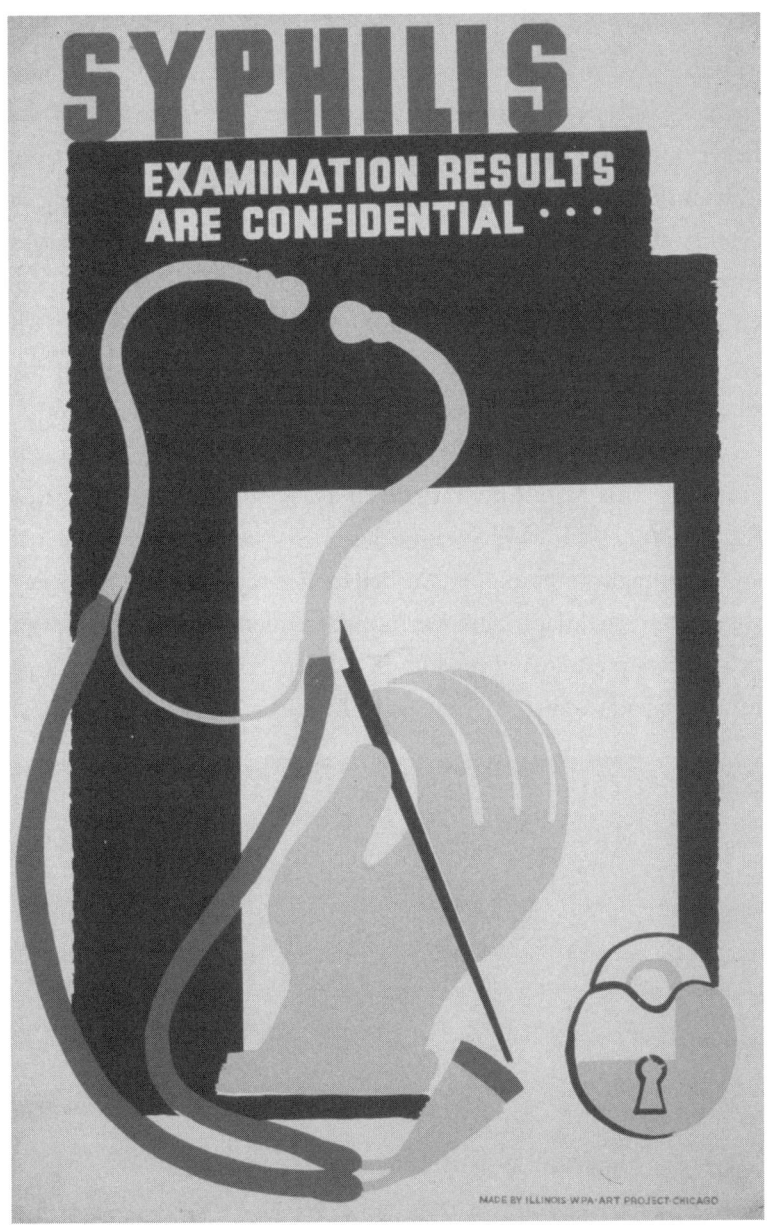

Public health poster from the Illinois WPA Art Project, 1936–40.
Here the stethoscope appears as a key element in a chain of
professional medical communication, particularly confidentiality.

block (where there is a delay or blockage affecting the electrical impulses that make the heart beat) and atrial septal defects (where there is a hole in the wall separating the two upper chambers of the heart, meaning blood leaks between them). By timing systolic murmurs in relation to the second heart sound he was also able to classify them into 'ejection-systolic' or 'pansystolic' murmurs. As John Camm, Professor of Clinical Cardiology at St George's Hospital, writes in Leatham's *Lancet* obituary: 'he turned what had been an art into a science . . . he was a very fine exponent of the use of the stethoscope, which he honed through constant experimentation.'[5]

Innovation in the design of acoustic stethoscopes continued alongside this experimentation. In 1926, for instance, Maurice Rappaport and Howard Sprague designed a stethoscope with a bell and diaphragm combination, the bell for listening to lower frequency sounds and the diaphragm for higher frequency ones. This design proved very popular and versions of it are still made and sold in large numbers today. In 1961 Dr David Littman, a Harvard Medical

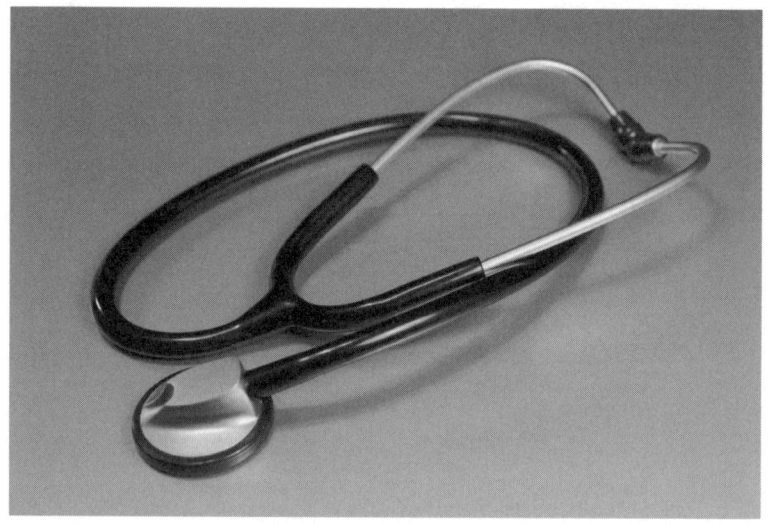

Modern stethoscope.

School professor, created a stethoscope that was similar in appearance to the Rappaport-Sprague but lighter and with improved acoustics. The Littman design and brand remains popular.

As the twentieth century progressed, then, the introduction of a range of new medical technologies and medicines vastly enhanced the diagnostic and curative efficacy of medical professionals. The stethoscope was used by, and symbolized, an increasingly powerful and high-status profession.

Nurses and the Stethoscope

During the twentieth century, associations were developed and consolidated between the stethoscope and fields and professions beyond just the doctor. Nurses began to use the instrument in the 1930s, usually to measure blood pressure using a sphygmomanometer.[6] The sphygmomanometer combined an arm cuff, pump and a dial that displayed a measurement of pressure expressed in millimetres of mercury. Taking a patient's blood pressure required the use of a stethoscope, with the user listening over the brachial artery (the major blood vessel of the upper arm) at the antecubital fossa (the inside of the elbow). The stethoscope enables the detection of the so-called Korotkoff sounds, named after the Russian surgeon Nikolai Korotkoff, who first observed them in 1905.

When taking a blood pressure reading using a sphygmomanometer and stethoscope, the cuff is positioned around the upper arm and inflated using the pump until the blood flow through the artery is completely cut off. Because there is no flow of blood, nothing is heard through the stethoscope, but when the cuff is gradually deflated a thumping sound can be heard at the point where the blood pressure is sufficiently high to begin pushing blood through the narrowed artery. When this thumping sound begins, a reading can be taken from the dial which equates to the maximum pressure of the blood in the artery (the so-called systolic blood pressure). The thumping sounds continue as the pressure in the cuff falls, and then cease at the point when the cuff no longer exerts any restricting force on the flow of blood and there is no turbulence. The flow has become smooth, and hence silent. The reading on the dial at this point gives the lowest pressure of the blood in the artery (called the diastolic blood pressure). Blood pressure readings are expressed as the systolic reading 'over' the diastolic one. Taking a blood reading requires the close coordination of auditory and visual sensing, as well as the capacity to manipulate the stethoscope and control the pressure in the cuff. Many doctors and nurses still use this technique, and it continues to be taught in medical schools, although in Western settings machines are often used that have inbuilt sensors and display readings digitally.

By the 1960s, nurses' use of stethoscopes was well established. Manufacturers began to market instruments directly to nurses, but they sometimes branded them in what we might regard today as patronizing terms, such as the 'assistoscope' or the 'nursescope'.[7] By the 1970s some nurses were also using stethoscopes as part of abdominal and chest assessments. In 1972 David Littmann, inventor of the popular Littmann Stethoscope, wrote an article in the *American Journal of Nursing* that detailed how to listen for the heart sounds and mentioned a variety of lung sounds and their meanings that were relevant for nurses. He stated that 'the stethoscope is as much a part of the nurse's armamentarium as the doctor's.'[8] Although he had sold his stethoscope company in 1967, Dr Littmann still worked as a consultant for the new owners, so perhaps he had some interest in opening up the new market in nurses' stethoscopes.

In medical practice today nurses use their stethoscopes for a wide variety of procedures, and the instrument is a standard piece of nursing gear. Some patients, however, still hold the idea that the

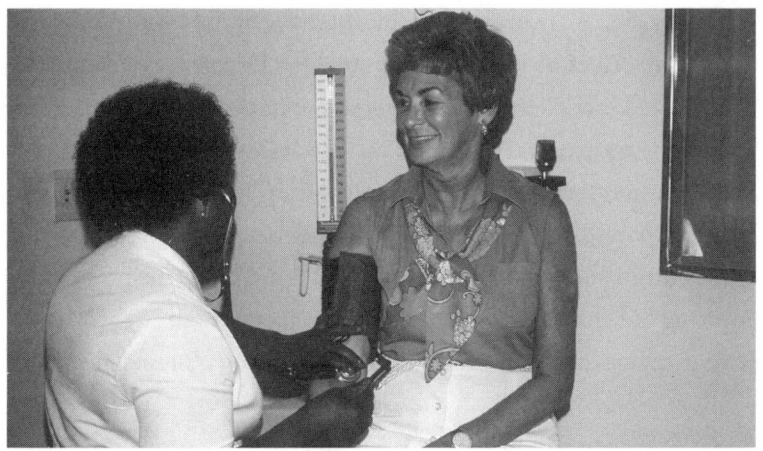

Nurse conducting a blood pressure examination using
a sphygmomanometer and a stethoscope, 1980.

bearer of a stethoscope must be a doctor. For instance, during his fieldwork, Tom interviewed a consultant nurse who ran specialist clinics in the Cardiothoracic Unit at St Thomas' Hospital in London. During the interview she explained:

> I work as a nurse in a clinic, but I don't wear a uniform, and I have a stethoscope. Now, patients in the heart failure clinic I see time and time again, they get used to the idea that I'm a nurse and I examine their chest and other things, but patients in the chest pain clinic who I only ever see once ... I always introduce myself as a nurse. 'I'm Elaine, I'm the *nurse*,' and I emphasize the word and I say, 'Your *doctor* has asked me to see you.' Now what I see a lot of them doing is looking at my badge throughout the consultation, so you'll be chatting to them and their eyes kind of go like this ... [staring at Tom's ID badge] and you can guarantee that 80 to 90 per cent of the time, when you say goodbye to them, they say 'Goodbye, doctor.'

In an article celebrating the bicentenary of the invention of the stethoscope, doctors Liliana David and Dan Dumitrascu refer to the 2015 Miss America contest. In one phase of the competition, Miss Colorado appeared wearing nurse's scrubs and sporting a stethoscope. Some commentators on the event suggested her outfit was inaccurate and inappropriate, sparking a lively debate as to whether or not nurses do in fact wear stethoscopes. Many nurses responded by pointing out that they frequently use the instrument in their work, posting images of themselves and their stethoscopes on social media.[9]

Non-Medical Uses

Nurses are not the only ones to have appropriated the stethoscope. Although the focus of this book is on the stethoscope as used in medicine, the instrument has been applied in a variety of other domains. At the same time, human bodies have not been the only objects of stethoscopic scrutiny.

Mechanics have long used the stethoscope to detect and analyse problems in cars.[10] In the interwar period, the German car mechanics trade was being shaped into a craft profession and mechanics drew on comparisons with medicine. Mechanics described cars metaphorically as patients, and their technical problems as diseases. From the 1920s onwards, pictures of mechanics posing and looking like doctors were common, and medical techniques and terminology began to be used. In their historical study of this development, Stefan Krebs and Melissa Van Drie quote from a mechanics trade journal from 1926: 'if the physician cannot make his diagnosis by the appearance of the patient, he will take his stethoscope and listen to the patient's body. This is how you ought to proceed with the car engine as well.'[11]

The mechanic's stethoscope, like the doctor's, started as a monaural listening tube, and developed into a binaural version. As in medicine, listening with these instruments was hard, the difficulty lying in telling apart the sounds of the different parts of the car and knowing how sounds corresponded to different mechanical problems. There was no formal training for this. Mechanics learned how to listen as part of an apprenticeship system that is common for many trades in Germany. As was to be the case in human medicine, the stethoscope in the mechanics trade was later threatened by instruments which promised more objective, often visual results, such as pressure gauges and oscilloscopes. Visual imagery of car mechanics in magazines and in advertisements reflected this shift, and if they showed mechanics dressed like doctors, they also tended to include more sophisticated technologies than the stethoscope. However, there were many mechanics who resisted these changes, continuing to listen diagnostically as an important part of their professional identity and a way of gaining the trust of customers.[12]

Motor mechanic listening to a car engine with a stethoscope, 1927.

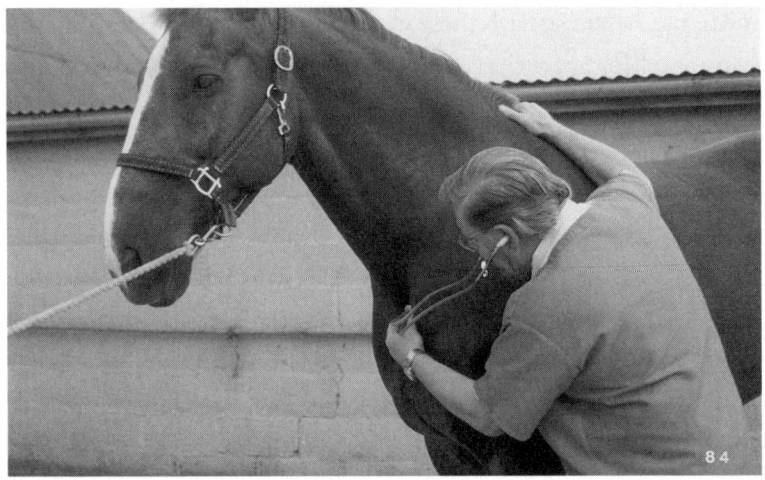

A stethoscope being used on a horse.

Like engines, animal bodies also became the recipients of stethoscopic attention. In fact, the potential for the use of the stethoscope in veterinary medicine was foreseen by its inventor. Laennec suffered from poor health, but he recognized the restorative value of breaks in the countryside, where he could enjoy fresh air, wholesome food and vigorous exercise. One of the pursuits he found most beneficial was hunting. Evidently, however, 'even in the woods, stethoscopic preoccupations were not far from his mind.'[13] During one hunting expedition, while he and his hunting partner were resting beside a ditch, Laennec grabbed his companion's dog and began vigorously percussing and auscultating it. Perhaps it was this incident that led Laennec to reflect on the possibilities for the use of mediate auscultation in veterinary medicine.

Laennec believed that mediate auscultation would not be as useful to the same extent in animals as it was in humans:

In the first place, in them we lose at once all the signs supplied by the voice. But there are likewise many other obstacles to

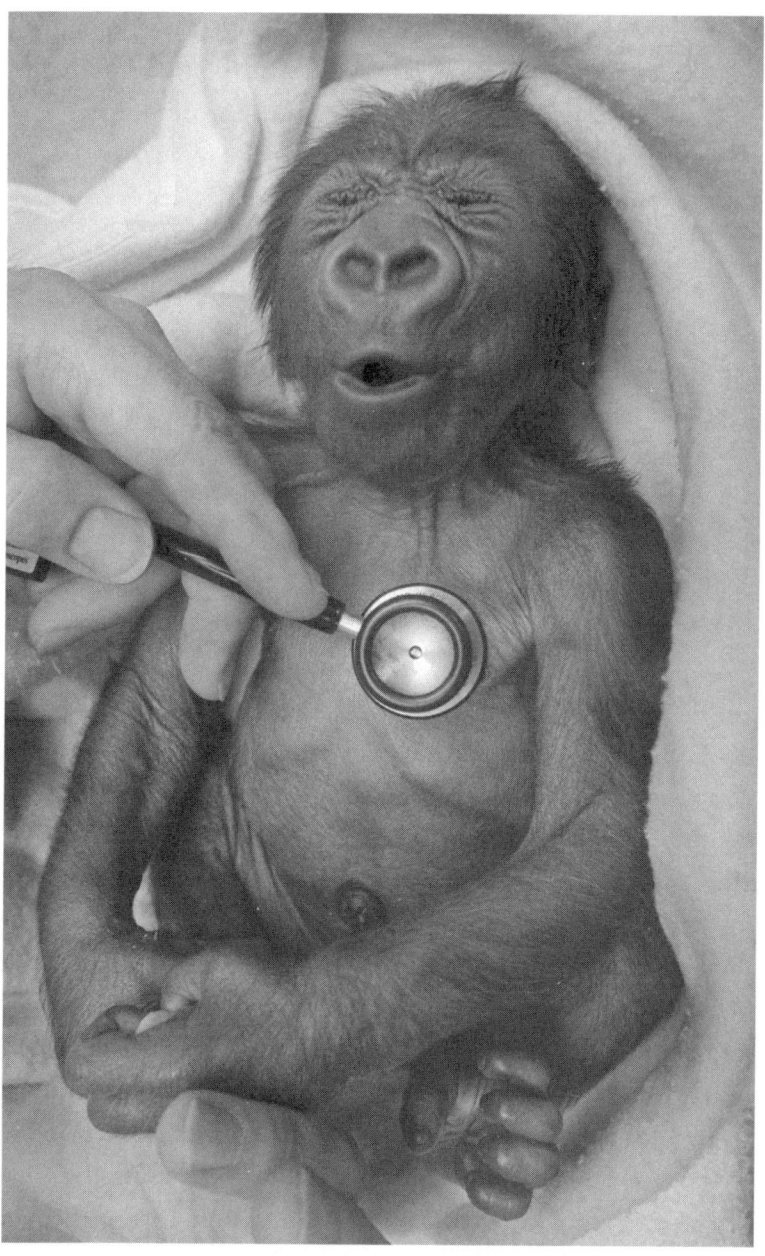

Stethoscope being used on 'Yakini', a baby lowland gorilla,
during a check-up at Melbourne Zoo, 1999.

the use of the stethoscope in animals. In the larger quad-
rupeds, such as the horse or bullock, the exploration of the
heart becomes extremely difficult, on account of the inconvenient posture necessary to attain it, and on account of
the form of the sternum. In the horse, and probably in all
herbivorous animals, the respiration is very indistinct, being
indeed hardly audible even when the animal has just ceased
running. I am, however, of the opinion, that in the state of
disease it would be more perceptible in the sound portions
of the lung, the action of which is, in such case, doubled or
tripled; and accordingly, I found, in one case, that it was easy
to recognise peripneumony in a cow, as in the human subject.
I ought to add, that my researches on auscultation in the
diseases of animals have been very limited, but I am still of
opinion, that it will be found very useful in such cases, more
especially when conjoined with percussion.[14]

Just as in human medicine, the stethoscope took some time to
be accepted into veterinary practice.[15] In the nineteenth century,
for instance, the British veterinary surgeon William Youatt (1776–
1847) insisted that the stethoscope was an inconvenient instrument
for vets, but admitted that it had alerted him to the importance of
sound, recommending 'the application of the ear to the chest and
belly of various animals'.[16] This procedure enabled the identification of a cow's pregnancy, for example, because 'the beating of the
heart of the calf will be distinctly heard' as would 'the audible
rushing of the blood through the vessels of the placenta'.[17] The
introduction of flexible designs of the stethoscope meant that some
of the inconvenience identified by both Laennec and Youatt could
be avoided, and in the twentieth century the instrument became a
standard piece of veterinary medical technology. It is used routinely,

and is considered to be as valuable in the diagnosis of, for instance, heart and lung disease in animals as it is in humans.

Stethoscopic listening has also found application in times of war and conflict, not only in checking the health of new recruits, serving soldiers and casualties, but in detecting vibrations and identifying sounds which might indicate the presence of the enemy. During the First World War in the French village of Givenchy, both sides engaged extensively in the practice of digging tunnels under enemy positions in order to lay mines that could later be detonated from a safe distance. When Allied digging teams paused in their work 'a man was always left at the face of the shaft with a stethoscope, listening for Germans working. While they could be heard it was known to be safe; only when they stopped was the detonation of a German charge likely.'[18]

French sappers using geophones to listen for sounds of Germans mining or countermining under French trenches. *Scientific American*, CXVI/23 (9 June 1917).

The First World War also saw the introduction of the geophone (combining the Greek words for 'earth' and 'sound'), which 'consisted of a pair of wooden discs about four inches in diameter by an inch and a half thick, in the centre of which was a layer of mercury contained between mica plates: these were connected by rubber tubes to stethoscopic earpieces.'[19] The device was said to magnify sounds by two-and-a-half times. Placing the two discs in contact with the ground and with earpieces inserted, the listener would move one disc in an arc round the other until any heard sound was of equal intensity in each ear. The origin of the sound was then located in the line at right angles to that between the centres of the two discs. Skilled listeners could recognize different activities and estimate how far away, and how deep, enemy tunnels were.

A depiction of geophone use, 1910–30.

Listening for signs of the enemy in Vietnam, 1968, photograph by John Olson.

The stethoscope continued to have applications in later conflicts. It was carried by bomb disposal experts in the Second World War, allowing them to hear the ticking of the clocks or timers on the bombs and to gauge whether or not they were still active. Stethoscopes were also used by the U.S. military during the Vietnam War as part of efforts to detect enemy troops.

Beyond war zones, the stethoscope has been used by plumbers and engineers to detect leaks in concealed or buried pipes. The instrument is also associated with eavesdropping or listening in, as the cartoon from *Punch* below suggests, on the voices and movements of those deemed (in this case romantically) 'interesting'. Wherever attentive listening offers a means of investigating things that are not directly visible, the stethoscope can be pressed into use. The instrument can sometimes enable surreptitious and secretive forms of listening which may carry an illicit thrill or are associated with tension or danger. Films depict thieves cracking open safes, for example, with the help of a stethoscope. In a short story in the literary magazine *Meanjin*, the Melbourne-based radio producer Dominic

SCIENCE APPLIED TO ART.

ANGELINA SQUILLS (THE DOCTOR'S DAUGHTER) BY A JUDICIOUS USE OF HER FATHER'S STETHOSCOPE, IS ABLE TO DETECT AND ENJOY THE DELICATE TENOR VOICE OF THE INTERESTING YOUNG CURATE WHO LODGES NEXT DOOR.

'Science Applied to Art. Angelina Squills (the Doctor's daughter) by a judicious use of her father's stethoscope, is able to detect and enjoy the delicate tenor voice of the interesting young curate who lodges next door.' Illustration from *Punch; or, The London Charivari*, LX (13 May 1871).

Gordon lives out his own fantasies of cracking a safe, stealing a stethoscope from his doctor to open the one in his local bowling club. The end result shows how this may just be cinematic theatrics:

> I figured this could be my big moment, it could be where I acquire a stethoscope and open the safe like I'd just seen in some film (I can't remember which). But I had to get a stethoscope. I made an appointment with a doctor ... I saw an empty room and I ran in, saw the stethoscope on the desk and grabbed it ... So here was my big moment, the moment where the thief uses tools and knowledge of equipment to get into a safe. My hands were shaking as I put the stethoscope up to the safe ... I turned the tumbler [part of the safe's lock mechanism] and listened, nothing. I turned and turned,

but still nothing. No clicks. Where was the sound? Was I doing it wrong? I stayed there for ages, sitting on the floor with the stethoscope, trying to find the pulse of the safe.[20]

At the same time, just as the very first stethoscope was improvised from a stack of paper rolled into a cylinder, doctors have noted that it is sometimes possible to make do, for example, with a wine glass or vacuum flask if no stethoscope is available. Bishop notes that 'Glasses or tumblers are sometimes portrayed in plays and films, being used to listen in on conversations in adjoining rooms.'[21] All kinds of objects, then, have stethoscopic potential.

Despite its varied applications beyond medicine, the stethoscope is still firmly associated with the doctor. Toy versions in doctors' kits socialize children into the practices of medicine and help to consolidate the centrality of the stethoscope to medical identity. Some children's stories have the same effect, for instance Julia Donaldson's *Zog* (2010), in which the princess who allows herself to be captured by the eponymous young dragon explains that really she wants to be a doctor: 'Travelling here and there/ Listening to people's chests/ And giving them my care.' The knight who arrives to attempt to rescue her shares the same dream. Removing his helmet, he exclaims that he would rather wear a 'nice twisty stethoscope'.[22] Today the stethoscope is also a symbol of not only doctors but medical knowledge more broadly. We see it used widely in posters for medical services, while Wikipedia uses the image of a stethoscope with a cross through it to denote that an article does not offer medical advice. In the following chapter we see how wearing, handling and using the stethoscope remains an important part of performing medical identity for medical students today. Getting to grips with the stethoscope, however, is not a straightforward task. It never has been.

5

Learning

The stethoscope had implications for the body of the doctor. During the mid-nineteenth century, auscultation began to be taught in medical schools across Europe and throughout the United States, and doctors were increasingly expected to be familiar with the technique. While it was one thing to own a stethoscope and to have tried it out, gaining real mastery of the instrument involved developing the capacity to detect and distinguish a wide variety of different sounds, as well as learning the diagnostic significance of each. The stethoscope not only repositioned the patient body as a dynamic acoustic space amenable to interpretation but reconfigured the perceptual skills of medical students and young doctors, requiring in particular that they take on new auditory sensitivities and capabilities. Mediate auscultation involved 'the coproduction of an organized universe of analyzable auditory signs and an auditorally sensitive body that was able to decipher their meanings'.[1] In this chapter we explore the creation of this sensitive body.

Soon after they enter medical school, and sometimes even beforehand, most medical students acquire their own stethoscopes. They often display them proudly around their necks and across their shoulders as symbols of their intended profession (much to the dismay of some doctors who find this practice pretentious, unprofessional and unhygienic).[2] During our fieldwork we found that some medical students reported feeling like pretenders when wearing their stethoscopes for the first time. They were aware of the gravity of donning the symbol of the doctor and of the knowledge and skill associated with them wearing the instrument, but equally knew that they had no idea what to actually do with the stethoscope. Tom was also conscious of feeling like a fraud during his fieldwork when he wore a stethoscope. He was still just learning to use it, and the fact that he wasn't a doctor or even a medical student made it feel like a double deception. Anna had a similar feeling about the stethoscope after she stopped practising medicine and became a full-time anthropologist. She was even reluctant to listen to patients' chests during fieldwork when invited. These acts felt somewhat fraudulent considering she no longer practised in the profession. Revealingly, however, this sense of being a 'fake' also in its own way confirms how important stethoscopic experience and expertise is to our notions of what constitutes a 'genuine' doctor.

As a medical student in London, Ontario, rheumatologist Pari Basharat penned a poem entitled 'The Stethoscope'.[3] She describes herself as nervous and conscious of how little she knows walking into a hospital room, 'The last in a long line of observers', not knowing what to expect.

I hear him first.
His breathing is heavy, laboured,
So loud it almost drowns out
The droning of the machines,
The humming of the fluorescent lights.

The man in this room is dying. There is perhaps something stethoscopic in Basharat's 'hearing first', and the listening continues, though once again it is unassured and uncertain:

> I glance at the monitors surrounding him,
> Listen with the ears of a novice
> As the more experienced physicians interpret the fuzzy
>   screen images,
> Examine his chest,
> Confer with one another.

Basharat imagines herself as 'a transient figure' in 'the final chapter' of the patient's life: 'Merely observing, with my limited medical knowledge,/ Barely able to grasp the jargon of the conversation going on around me.' She tries to think beyond the man's symptoms and imagines how his family will react when they learn that there is nothing the medics can do. Soon Basharat's thoughts are interrupted. It is time for the group of which she is a member to move on:

> 'Nice meeting you' I say . . .
> Hollow, ridiculous-sounding words.
> In the hallway the case is explained . . .
> I imagine him inside,
> Alone,
> As we discuss his fate.
> We then walk hurriedly down the hall . . .
> My stethoscope slung snugly around my neck,
> A false semblance of confidence, assurance . . .
> On to the next doorway.

It is interesting again how Basharat makes herself sound like a novice auscultator who is 'barely able to grasp the jargon' that is being used and how her own words are 'hollow' and 'ridiculous-sounding'. The stethoscope is something for Basharat to hide behind. The fact that it fits her 'snugly' suggests she takes some comfort or even protection from it, but the instrument is also a mask, a prop in a performance of experience and competence.

Medical students, then, need to find a way for the stethoscope to become part of their diagnostic toolkit and an extension of their own bodies, their system of clinical sensing. This process of incorporation is not easy. Since its discovery, auscultation has often been described as difficult and time-consuming to learn. The difficulties involved in training novices in the technique were at the root of some objections to it and were a reason for its rejection by parts of the medical establishment in the nineteenth century. They continue to present a serious challenge to the status of stethoscopy in contemporary medicine. Despite these challenges, and perhaps because of them, educators in medicine have over time developed various ways of teaching auscultation, using a range of materials, linguistic devices and technological innovations, as well as their own bodies and those of their students.

This chapter focuses on the intricacies of learning how to use a stethoscope, examining styles and methods of teaching auscultation. While we have documented the stethoscope's spread in a rather linear fashion through particular historical characters and places, the techniques and materials involved in teaching and learning auscultation have telescoped through historical periods, so that practitioners in some medical schools might still use the blackboard to teach lung and heart sounds, while in others they are using digital set-ups. In order to attend to this variety in teaching methods where the old and new intermingle, we focus on spaces and tools of learning, particularly lecture halls, textbooks, wards and audiovisuals. We suggest that each site and each technological apparatus is part of an attempt to produce what Anna and Melissa Van Drie refer to as 'sonic alignment'.[4] This is a state where the trainer and the novice attempt to hear the same thing and achieve a kind of intersubjective experience so that the sounds of stethoscopy can travel across generations of learners.

Lecturing and Lecture Halls

The lecture hall could be seen as a very traditional site of learning. It is also a critical site of both continuity and change in education and offers an excellent starting point for looking at how auscultation has been taught over time.[5] While criticized by some in educational practice as redundant in the face of online or small group learning, in many places around the world the lecture continues to be a place where medical students 'learn together'.[6]

Lecturing about auscultation was an important element in the dissemination of the practice in the nineteenth century. In the 1830s, doctors such as James Hope at St George's Hospital in London taught stethoscopic technique using diagrams, alongside verbal instruction, during 'experimental sessions'. As described in the *London Medical Gazette*, as late as 1837 some doctors were still giving lectures which minimized the importance of the stethoscope in clinical diagnosis. Hope, however, was one of the doctors in England working to turn the tide of scepticism.[7] He evidently relished teaching auscultation and at the end of his lectures for medical students would hold auscultation competitions, a practice which some medical educators continue to this day. He was highly regarded as a stethoscopic listener himself and had been awarded a prize for proficiency in auscultation.[8]

Hope realized that it was partly the fear of learning the stethoscope that was preventing its uptake, and so his public lectures focused on how easy it was to use. His sessions involved experiments, as described by one member of his audience:

> Dr. Hope took four students, all novices in auscultation, and as several of them did not know the sound of valvular murmur, he introduced a single patient to afford them the opportunity of hearing it. He then ascertained by examination, that they were acquainted with the anatomy of the heart, and with its situation and relation to the exterior. This being done, he occupied ten minutes in giving an explanation, elucidated by a chalk diagram, of the mode of discriminating between the various valvular diseases and in catechising to ascertain that it was understood.
>
> Six patients presenting five distinct varieties of valvular disease, some complicated and some obscure, were now introduced and each pupil examined as many of them as the leisure of the patients would permit, writing his notes and diagnosis on the slips of paper ... Out of sixteen diagnoses which were made, one alone was partially defective.[9]

Several medical educators published books of their lectures on auscultation. Peter Mere Latham (1789–1875), a highly regarded physician and teacher who had been quick to adopt the stethoscope, published his *Lectures on Subjects Connected with Clinical Medicine* in 1845 and subsequently acquired the nickname 'Heart Latham'. He is regarded as one of the early pioneers of cardiology.[10] Another nineteenth-century physician, John Hughes Bennett (1812–1875) of the University of Edinburgh, published a book of his lectures, including some on auscultation, which appeared in multiple editions in the UK and the United States and was translated into Russian and Hindi.[11] The book details the general rules to be followed in the practice of auscultation, explains what to look out for in pulmonary or cardiac auscultation and offers hints on how to hear sounds on a healthy body, potentially the student's own, that they might also find in pathological cases. The commercial success of Bennett's published lectures raises questions as to whether this 'text book' form was filling a void, and perhaps sparked demand for more lecturing on auscultation. In Bennett's preface he states his lectures were published 'to facilitate the studies of gentlemen attending the Author's clinical instructions in the Clinical Wards of the Royal Infirmary of Edinburgh', but notes that he has imagined they may be useful to many in other schools of medicine.[12]

New technologies entered the lecture hall space in the early twentieth century, such as the electronic stethoscope, enabling multiple students to listen together to the same sound.[13] One example of this was the stethophone, a device based on telephone technology, which was first installed at Massachusetts General Hospital in the United States in 1923. The mouthpiece was placed on the patient in the hospital to record their sounds, and these were then transmitted to the nearby lecture hall via a vacuum tube amplifier and distribution system. Up to 125 students were then able to listen to the patient live at their seats by applying their own individual stethoscopes to this mechanism. The elaborate nature of this set-up is testament to the value that was placed on learning to listen to heart sounds at the time.

Importantly, the lecture hall offers a theatrical space for training in the detection of cardiac and other bodily sounds.[14] Recounting an introductory lecture given by Dr W. Proctor Harvey in the 1960s,[15] Van Drie describes how when students entered the lecture hall they were greeted not by heart sounds, as they might have expected, but a recording of one of Beethoven's symphonies playing over the lecture hall speakers. The professor then asked his students: 'Who heard the violin? Who heard the kettle drum? Who heard the French horn?' Dr Harvey used this exercise as a way of helping students to 'tune in' to the sounds they would later hear on patients through their stethoscopes, but it was also clearly his way of adding drama and an element of the unexpected into what students might have anticipated would be a more mundane educational experience. There is a distinct element of showmanship here, as indeed there is to all good lectures.

In contemporary medical schools, a range of learning aids may be used in the lecture hall. Blackboards and chalk, for example, first introduced into educational settings in the nineteenth century (when lecturing moved from being based on reading to incorporating question-and-answer exchanges), can still be found in classrooms and lecture theatres around the world, though blackboards are becoming obsolete in many medical schools, replaced by whiteboards, smartboards and multimedia presentations.[16] In her work on the multisensorality of chalkboard learning in medicine, the anthropologist Rachel Vaden Allison suggests that the sonic and tactile nature of chalkboard sketching affords an anatomical imaginary that correlates closely with the multisensory way in which medical students learn about bodies in the dissection labs and on hospital wards.[17]

In many medical schools today a lecture on heart sounds might start with the teacher, often a clinician, introducing what are essentially abstract principles of auscultation. This might entail, for example, describing the anatomy of the heart or trying to visualize the heart sounds through sketches or diagrams. These may be pictures showing waveforms or graphic representations of abnormal beats. Visualization has long accompanied the spread of stethoscopic listening and the teaching of sonic skills in medical and other

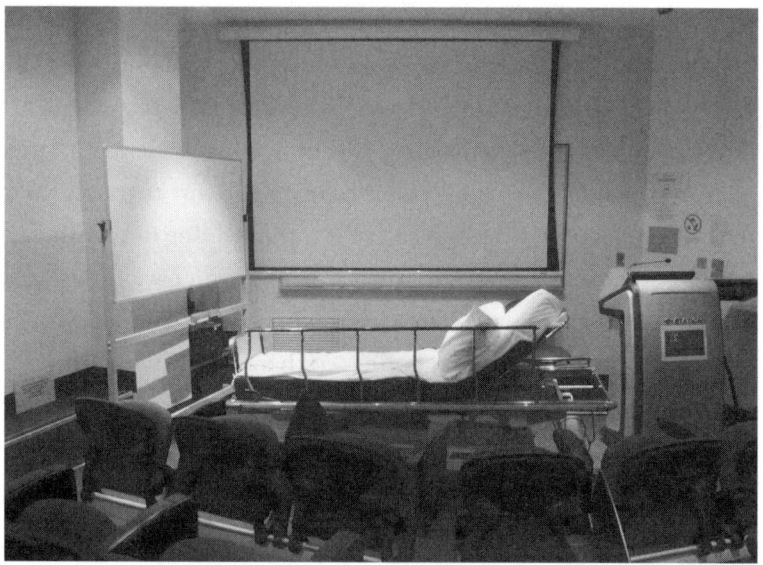

A lecture theatre in the Royal Melbourne Hospital.

scientific settings.[18] These sketches are a way of rendering and sharing sonic information with novices, aiding sonic alignment through a form of sensory triangulation. Van Drie's work highlighted the importance of visualizations in textbooks over the years, showing that writers increasingly privileged images over text when training skills in auscultation, gradually moving to multimedia material.[19]

Alongside sketches and drawings, the lecturer might also give a personal performance imitating a murmur, and the students may be able to listen to some recorded sounds embedded in a multimedia slide presentation. The students make notes but also have their own smartphones and other electronic equipment to record the lecture. Importantly, many lectures are available online. Medical students watching these videos are no longer just listening in a lecture hall with others but may be in their bedrooms, at the gym, on the train and so forth. This has been another stage in the evolution of the lecture as a pedagogical form. Indeed, thanks to digital platforms and methods of distribution, lectures on the stethoscope are perhaps being consumed more widely than ever before.

Lectures about stethoscopy still generally involve a bodily form of storytelling.[20] As well as referring to drawings, recordings and videos, a cardiologist, for example, might well use gesticulations and mimicry to teach cardiac sounds to medical students.[21] Auscultation is a practical skill, and neither a large lecture nor a video can provide the much-needed interaction with patients in the flesh that is so essential to learning the technique. Sometimes a patient will be invited into a lecture, a practice that has been adopted in medical schools for centuries and which means that everyone present can try listening, grasping the practical elements of handling and placing a stethoscope, managing the interaction with the patient, and also trying to hone the sensory and cognitive elements of detecting and recognizing a murmur. Although this can take place in lectures, it mostly complements the strong tradition of bedside teaching in medical training.

Bedside Teaching

The mode of learning that most medical students look forward to happens on the wards, with patients. In the second chapter we documented how the rise of the stethoscope correlated with the rise of bedside teaching in the nineteenth century, the practice initially being more widely accepted in European centres of medical education such as Paris and Edinburgh, and less so in others such as London. Foucault suggests that seeing and doing at the bedside under the clinician's expert guidance emerged as a form of medical training in the late eighteenth century.[22] Much-used quotes such as 'to study the phenomena of disease without books is to sail an uncharted sea whilst to study books without patients is not to go to sea at all,' and 'Medicine is learned by the bedside and not in the classroom,' are traced to William Osler, whose skills in bedside teaching were renowned.[23] They convey the importance that became attached to learning at the bedside.

For a large part of the twentieth century, depending on the medical school, medical students encountered patients either in their first year, as is the case in Australian medical schools, for instance, or after a few years of pre-clinical training, as is the case in British ones. The traditional method of bedside teaching involves a small group of medical students and a teaching doctor visiting a patient at their bedside. Nowadays, the bedside is much less common as a teaching site, a development which is often attributed to consultants having limited time on the wards, thus reducing the time for teaching; patients not staying as long in hospital, meaning there is less chance for students to see them during their stay; and increased reliance on sites of simulation and training skills outside of the hospital where it is considered more safe, more standardized and more efficient to learn. Many medical educators complain that 'the long tradition of bedside teaching has now become more of a relic than a norm, to the disadvantage of both the current generation of medical trainees, and to the patients too.'[24]

Both authors were fortunate to have witnessed the teaching of auscultation at the bedside during their fieldwork. This generally involved an organized tutorial with a professor who had pre-selected patients on the wards with 'interesting signs', whom they then visited with a tutorial group. Sometimes the tutorials were taken by retired doctors who still held instruction in clinical skills dear. Anna spent time with Professor Smith (as per anthropological convention, this is a pseudonym) in the Melbourne hospital where he worked. He would meet his students on the ward smartly dressed in a white lab coat, with index cards in his pocket to make notes or drawings. He was kind and imposing, and the patients and students alike were fond of him. During the course of an hour the class would visit three or four patients. Each time a student would be chosen to conduct a physical examination in which taking out the stethoscope always seemed to be the climactic gesture. The other students waited, huddled around the patient's bed, to see what their classmates heard. Often the student would attempt to describe the sounds, or in other cases, they would not hear them at all. The others would then be called on to listen. Finally, the professor would listen, too, and declare what the sounds were, describing them and asking questions about their related pathology.

Sometimes teachers find a sound in a patient's chest that can be described as a 'classic' murmur. It may be easy to find, loud, typical or rare. These sounds are very important in teaching students auscultation, and patients with 'beautiful murmurs' are often listened to many times during their hospital stay.[25] Patients react differently to this attention. Some are curious or lonely and enjoy the extra attention they receive from the students; others see offering themselves up to be listened to as reciprocity for the medical care they have received and for which they are grateful. Others resent the intrusion, and feel objectified. This is an exchange that undergirded the provision of care for the sick poor for centuries and which is still very much in evidence today.

At other times in Anna's fieldwork, bedside teaching happened in the midst of the everyday work of the ward, with medical students trailing behind medical teams as they visited patients. In such cases students were often tasked with writing in the clinical notes of a patient, including drawing visualizations of what had been heard with the stethoscope by the doctor who examined the patient. The doctor might declare that they had heard 'crackles', for example, and the student was required to write this in the notes. These textual and visual renderings of the audible signs of the ward-round clinical examination help to both consolidate the sounds and render them meaningful to others. The students learn by copying those who have made similar diagrams in the clinical charts, though this can also lead to cases where diagrams are repeatedly copied without someone actually listening to the sounds themselves.

Textbooks and Audiovisual Materials

As well as attending lectures and heading to the wards to hear 'real' sounds, students learn auscultation from books and audiovisual material. Van Drie spent time in the archives of medical libraries and museums in Paris, London and Washington trying to understand how medical students in the past learned to listen through textbooks and sound recordings. Studying medical textbooks published between 1950 and 2010, she found that they constituted an important resource and offered guidelines so that medical students could learn to elicit sounds with a stethoscope as well as to interpret them.[26] Description of the technique of how to use the stethoscope, for example, might include how to position the earpieces so that they fit comfortably, how to deal with clothing and patients' chest hair, how to create a good sonic environment for careful listening and how to avoid awkward gestures and noises such as banging the stethoscope into the bedrails.

Textbooks are often re-edited and republished numerous times over, which gave Van Drie a way in which to understand how teaching auscultation has changed over time. The content of the chapters on teaching auscultation all involved explanatory text, drawings and photographs, diagrams and charts, but the space allocated to auscultation in the textbooks gradually diminished. First to disappear from the text were explanations of how stethoscopes worked and the latest in stethoscope design. The instrument lost its novelty, and its redesign was not regarded as important information for the medical student. The next set of details to go were the descriptions of sounds, which went from complicated classificatory systems to more simplified accounts. Gradually the descriptions of the technique of auscultation were also simplified into bullet points and diagrams.

Van Drie also noticed that in the 1970s there was an escalation in the production of audiovisual material to accompany the textbooks.[27] These materials worked to reproduce and represent the lung sounds and heart sounds that students would hear through their stethoscopes. They helped to facilitate collective teaching experiences. Although it was difficult to recreate the exact experience of listening to sounds with a stethoscope, producers of auscultation sound recordings went to great lengths to simulate a more complete bodily experience. For instance:

> In order to preserve realism this recording has been prepared such that faithful, life-like reproduction will be achieved only if you listen to the cassette with a stethoscope. Thus, the cassette player serves as a substitute chest and a stethoscope is interposed between the listener and the electronic 'chest' as it is in real life.[28]

The fragility of such a sensorial re-enactment is underlined by the precision needed in order to obtain the proper listening conditions:

> Hold the bell [of the stethoscope] 2 to 3 inches from the speaker of your tape recorder. If you place the bell on the speaker, you will hear more noise than breath sounds. If

you listen to the sounds without a stethoscope, they will sound unnaturally loud and booming.[29]

Like the textbooks, these recordings also changed over time, with gradually less description of techniques of how to use a stethoscope and more focus on listening to and visualizing the sounds themselves. Cassettes were soon replaced by CDs, and now, more commonly, links to websites of heart and lung sound repositories. There are a huge number of these. Students and doctors alike can follow tutorials, listen to sounds of diseased hearts and lungs, view illustrations and animations and test themselves with quizzes. Paradoxically, while giving students more and more opportunities to listen to body sounds, these resources also allow the medical listener in training to listen at a distance from patients. During Tom's fieldwork, medical students told him they would practise listening to heart sounds in the library, in their bedrooms and on public transport. One student told him about an interesting kind of cinematic experience he had had when listening to heart murmurs while on a bus across London. It gave traffic jams and other blockages the semblance of disease in the circulation system of the city.

Collections of heart sounds (gathered with a modified stethoscope) have also been adapted for artistic purposes. In Christian Boltanski's installation *Les Archives du coeur* (The Heart Archive), visitors to a small wooden house on the beach on Teshima, a remote island in Japan, are invited to record their heart sounds to add them to an existing archive which includes the heartbeats of people from Germany, Korea, Sweden, Italy and Finland, among other places. We might think of a body sound like the heartbeat as something evanescent, a trace or shadow that is present in one instant and gone

Christian Boltanski, *Les Archives du coeur*, 2010.

the next. But Boltanski uses stethoscope recordings in this piece to remake the heartbeat as something that endures in his extensive archive, amplified and shared in the exhibition, rather than being so intensely quiet and intimate that it is rarely heard even by the owner of the body that produces it. Boltanski creates an archive that immortalizes the heartbeats in the museum space, and while we might be inclined to think of one heartbeat as much like another, this installation suggests that a heartbeat may potentially be a distinct identifier, as much an aspect of a person's individuality as a visual feature captured by a portrait. In fact, the heartbeats stand in for people, making them present and absent at the same time. The archive of heart sounds played in the exhibition space means the hut also functions as a node where heartbeats, usually so geographically dispersed as people live and move around the globe, can be concentrated together.

Tutorials and Teaching Stethoscopes

Medical students now increasingly learn about auscultation in small group tutorial settings away from actual hospital wards. These tutorials can take many different forms. In Maastricht, first-year medical students learn about auscultation through dedicated tutorials on techniques such as the respiratory and cardiac examination. In a place called a Skills Laboratory dedicated to learning such skills, rooms are set up with beds meant to resemble those on the wards. Students follow a workbook which describes, like a recipe, the step-by-step method for how to conduct a respiratory or cardiac examination. When it comes to the stethoscope, students are instructed how to place the earpieces in their ears and told important sites of the body to listen to, such as at the trachea or over the lungs. The students are not meant to be finding pathological heart or lung sounds here, rather the focus is on learning the technique of using the stethoscope or practising percussion with a 'patient' lying or sitting on a bed, and how to tell the difference between the kinds of sounds they hear: the hollow percussion sound of lungs under the clavicle, for example, or the dull percussion sound of the liver.

Students still find auscultation very difficult to do, especially when it comes to listening to the sounds which are subtle and hard to detect. This is when an odd-looking instrument might appear in tutorials: the teaching stethoscope. This device conjures up images of many-headed figures from science fiction, but functions in much the same way as the communal listening that takes place with portable music players where people share headsets. The diaphragm is placed at the desired point on the patient's body while the teacher and student put in their respective earpieces. Like the stethophone which used telephone technology to broadcast a body sound from a

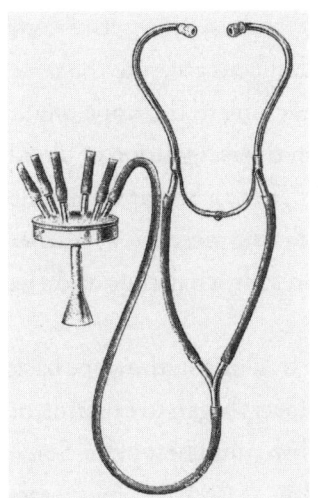 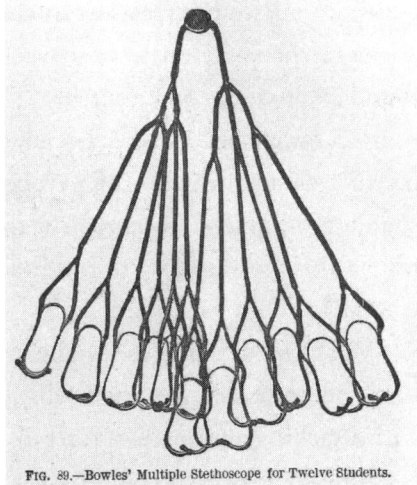

FIG. 89.—Bowles' Multiple Stethoscope for Twelve Students.

Teaching stethoscopes in print, designed by Dr William Evans, from *The Medical Annual* (1935); Bowles' multiple stethoscope, designed for twelve students to listen in, from Richard C. Cabot, *Physical Diagnosis*, 3rd edn (1905).

Doctors listening to an electronic heart recording through their own stethoscopes, 1920s.

patient to 125 students in a lecture theatre, the teaching stethoscope allows teachers and learners to have simultaneous access to the same sound and to focus on it together. The teaching stethoscope might be used, for instance, to coach students in the recognition of blood pressure sounds, which are often very soft. The teacher taps when the sounds grow quieter, accentuating their fading presence. A teacher might also use mimicry to try and represent what they and the student are both listening to.

Van Drie has traced the origins of the teaching stethoscope back to the mid-nineteenth century, when engineers began to experiment with attaching multiple earpieces to one listening chest piece. Some teaching stethoscopes could have as many as twelve earpiece sets attached to them, leading no doubt to a very daunting experience for the patients who had been chosen for listening practice!

Simulation

Although teaching in ward environments is still an important aspect of medical training, students increasingly learn and practise their auscultation skills away from the bedside, in simulated environments. A simulated environment for teaching clinical skills can entail anything from a room with one or more mannequins, to quite complicated multimedia digital set-ups designed to recreate a clinical context as closely as possible. Teachers and technicians can watch from behind one-way glass and manipulate the conditions encountered by students (including the sounds they hear) using hidden controls.

The teaching of medical students through simulation, particularly with mannequins, has a long history. Some date the first use of simulated patients to the 1900s,[30] however, examples of medical simulators made of hand-stitched leather and canvas can be found as early as the 1700s.[31] These models were highly interactive, some leaking fake body fluids. By the twentieth century, models had sophisticated sonic elements. 'Harvey', for example, a creation of researchers in the 1960s, was an elaborately engineered life-sized cardiology patient simulator that reproduced cardiovascular physical examination findings.[32] The model was named after Dr Harvey, who introduced symphony recordings into his lectures on auscultation. Nowadays, medical educators can buy mannequins which have been engineered to suit many different teaching purposes and be as haptically lifelike as possible. Some even include augmented reality apps for emotional engagement and interaction.[33]

The 75-year-old Japanese company Kyoto Kagaku sells various kinds of medical mannequins, and great care is taken to include interactive elements. For one model a doctor painstakingly recorded the sounds of many of her patients who had diseased hearts and lungs, as well as those who were classified as 'normal'. These recordings were then integrated so that when students applied their stethoscopes to the mannequin they could hear sounds in the areas of the chest which were anatomically and pathologically correct. The company is now attempting to create a more 'universal torso', with sound recordings made from patients not only in Japan but from U.S. clinics and hospitals. This effort to produce a universal torso highlights the ongoing quest on the part of medical education technology manufacturers to make tools which stretch across places (and markets). As humanities and social science scholars writing about medical practice have long argued, however, diversity in diagnostic signs is not simply a product of physiological and biological difference but an expression of historically particular modes of paying attention as well as the technological, spiritual, material and cultural environments in which perception takes place.[34] The universality for which the companies are striving, then, might not in fact be attainable.

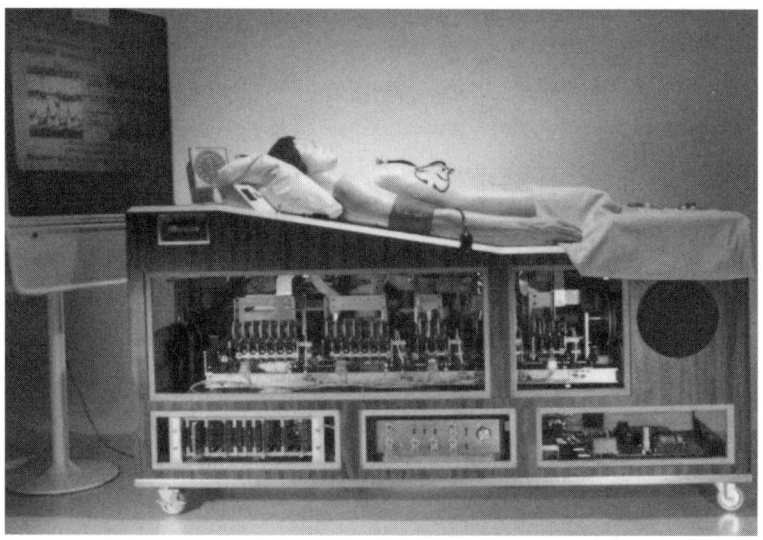

'Harvey – the Cardiology Patient Simulator', named after
Dr W. Proctor Harvey and developed by Dr Michael S. Gordon.

A typical scenario in which mannequins and stethoscopes may be used in a simulated environment is as follows. A group of students wait in a nearby room, one having been chosen to act in the role of a nurse. Students are then rushed into the simulation setting, the nurse saying that the patient is short of breath, they don't know what to do, please help. One student then performs the role of the doctor. As part of an examination, they might take the stethoscope out to have a listen. Through the plastic they can hear whatever those controlling the mannequin behind the one-way glass want them to hear – a wheeze or crackles, for instance – and the sounds are distributed through the mannequin's chest according to the clinical presentation. It is a form of teaching which many in Australia, the Netherlands and elsewhere advocate as offering students practical insights into clinical decision-making without any risk to the patient. Other teachers and educators are less certain of the efficacy of this form of teaching and the expense of these simulated environments.

Listening to Your Own Body

When learning to listen with the stethoscope, there is one resource that is cheap and always at hand, a living, breathing body with heart sounds and lung sounds ready for listening: the student's own body. One of the first bodies that students often listen to is their own, although this only became possible with the invention of flexible stethoscope tubing. The term used to describe the practice of listening to oneself is 'auto-auscultation'. Medical students generally do not have diseased hearts and lungs (although on occasion some do discover heart murmurs), but by listening to themselves they can develop a sense of what normal heart sounds are like and what happens when they breathe in or out and do other actions that affect their body's physiology. Their own bodies are at once a walking, fleshy textbook, a repository of body sounds and a practice patient.

This chapter has shown some of the range of strategies that help achieve a measure of stethoscopic 'sonic alignment' so that the skill of recognizing the sounds of disease can be passed on within communities and between generations of doctors. Auscultation is a time-consuming practice to develop and the challenges of sharing sounds between teacher and novice in reliable ways that create a consensus over what is heard have never been fully overcome. The time and effort involved in learning auscultation has historically put many doctors off, and in some places contributed to the slow uptake of the technique. This problem has not gone away, and can often count against auscultation when it comes to scheduling into busy curricula all the things that students must learn in order to operate in contemporary clinical contexts. In a medical environment where time and resources are tight, is teaching students to use stethoscopes still a sensible investment? This question reappears as a major theme in discourse on 'the death of the stethoscope', the idea that the instrument has had its day. It is to this discussion that we turn in the next chapter.

6

Obsolescence?

In an article for *The Atlantic*, 'Why Doctors Still Need Stethoscopes', the writer-doctor Andrew Bomback observes the different ways in which doctors have worn their stethoscopes in recent history.[1] The current generation, he writes, wears the stethoscope like a shawl draped over their shoulders, a style that has become iconic and is seen on television shows and advertisements. The way Bomback's father wore the stethoscope was different: the instrument hung from his neck but in such a way that the earpieces joined together at the back. Bomback writes that this method projects an image of the doctor just having used the stethoscope, or of having it at the ready, rather than it being a mere fashion accessory. Nowadays, Bomback writes, 'two centuries [after Laennec], a physician who tries to make a diagnosis based on the stethoscope exam is as anachronistic as a baseball scout who watches prospects rather than poring through all of their advanced statistics.'[2]

So is using the stethoscope a 'dying' art? This question has been hotly debated by doctors in hospitals, by medical school curriculum designers, by patients and also the popular press. Some doctors argue that auscultation is an intimate ritual that remains central to humanistic medical practice in the context of technological advance. Such arguments presuppose that the stethoscope is no longer a new technology as it was in the nineteenth century, but rather has become a 'traditional' medical practice.

Conversely there are those that argue that the stethoscope is far too user-dependent, the findings too subjective and the signs too difficult to interpret and share with others to be relevant today. Eric Topol, for example, a proponent of digital medicine, believes that artificial intelligence far outstrips the stethoscope. He writes in his book *Deep Medicine* (2019): 'Now that the stethoscope is 210 years old, and although it endures as the icon of medicine, it's time to reexamine the tools of the [physical] exam. The stethoscope is simply rubber tubing that doesn't record anything and, at best, can only transiently serve as a conduit for hearing body sounds.'[3] He goes on to write that he cannot share heart sounds with patients and that patients do not know what the sounds represent. He suggests that the smartphone visualizations that accompany ultrasounds are more readily comprehensible to the patient. These are becoming important factors to consider in healthcare contexts which are increasingly litiginous and image-based as well as patient-centred, and where care is sometimes conducted across vast distances.

In discussions as to the perceived relevance of auscultation, it is important to consider the kinds of technologies competing with the stethoscope over time and the ecology of practices within which the stethoscope sits at any moment. In this chapter we explore and expand on some of these 'competing' technologies and delve into other challenges that come with auscultation before turning to claims that despite these difficulties, the stethoscope should not disappear from the doctor's toolkit.

Competing Technologies

In 2013, queries about the relevance of the stethoscope were raised in an editorial in the journal *Global Heart*.[4] In particular, the editors questioned the relevance of auscultation in light of point-of-care ultrasound technology. Their comments led to an array of journal and newspaper headlines which predicted the stethoscope's demise:

Almost 200 years later, are we living in the final days of
  the stethoscope?

Stethoscopes Could Become Extinct, Doctors Say

Death of the Stethoscope

Trusty stethoscope faces threat from portable hi-tech[5]

The *Global Heart* piece predicted that the stethoscope would soon become a historical artefact:

Many experts have argued that ultrasound has become the stethoscope of the 21st century. While few studies have pitted ultrasound head-to-head against the stethoscope, there is evidence that ultrasound is more accurate even than chest X-ray in the detection of pneumothorax, pleural effusion, and perhaps even pneumonia. Ultrasound allows visualisation of cardiac valve function, contractility, and pericardial effusions with greater accuracy than listening with the stethoscope. And beyond the heart and lungs lie dozens of other organs and structures – well-described in the literature of point of care ultrasound – which are opaque to the abilities of the stethoscope.

The article continues: 'Certainly the stage is set for disruption; as LPs were replaced by cassettes, then CDs and mp3s, so too might the stethoscope yield to ultrasound.'[6]

But this is by no means the first time that doomsday predictions have been made about the stethoscope. A variety of technologies have promised clearer, more accurate, more objective results about the interior of patients' bodies. With each introduction of a new technology, the stethoscope has been subject to claims of obsolescence. For instance, X-rays, introduced in the late nineteenth century, were able to produce images of beating hearts and lungs during respiration that had never been seen before.[7] One enthusiastic Dutch physician wrote:

If one, watching the fluorescent screen, asks the patient to breathe deeply, it is indeed surprising how beautifully one can see the unfolding of the apex of the lungs, the movement of the diaphragm, and the partial covering of the heart by the inflating lung. There is no other means of investigation that offers us the possibility of such precise determination of the respiratory mobility of the boundaries of the lungs.[8]

Part and parcel of the claims being made for X-rays was their superiority to other diagnostic methods (principally auscultation) in the case of diagnosing certain illnesses. But while these claims were being made in conference halls in the early twentieth century, actual practice in clinics and hospitals did not always match. Pulmonary tuberculosis, for example, remained a disease that was clinically diagnosed with the stethoscope, in tandem with bacteriological examination of the sputum. Claims about the superiority of X-rays were rather statements as to the *promise* of the technology and the directions pioneering radiologists *hoped* it would take. One physician pushing for the X-ray expressed his dismay at his colleagues' reluctance to embrace the new technology: 'It is a pity . . . that elder members of the profession do not keep a more open mind on the subject,' he wrote in 1905. 'It comes to this: that what we can see with [our] eyes is often at variance with what in fancy we hear with our ears. In most cases our eyes are the safer guide.'[9]

Not everyone, then, accepted the new technology of X-rays. As Van Drie recounts in her discussion of clinical examination textbooks:

Prior and Silberstein [authors of the textbook *Physical Diagnosis: The History and Examination of the Patient*] open their chapter on the lungs by rejecting claims that the X-ray could replace the stethoscope. They state that both techniques provide unique diagnostic knowledge and should be used complementarily. They use a familiar cultural analogy to make their point: 'The situation is analogous to watching television – the picture without sound leaves much to be desired and the sound without the picture may be meaningless'. A friction rub, râles and wheezing cannot be seen on X-ray films and can be detected only by our senses. In fact, the findings on the X-ray film in many instances can be interpreted intelligently only when coupled with the physical findings. Consequently the chest roentgenogram will never replace the physical examination as performed by the skillful physician. Careful examination should enhance our ability to interpret the X-ray films, and the chest film should serve as a check on our physical examination (Prior and Silberstein [1959] 1963: 173).[10]

While a rearrangement of visual and auditory roles is allowed, complete erasure of auscultation, when skilfully performed, is not. As is still the case today, there are many who use the introduction of competing technologies to highlight the importance of the stethoscope.

Behind some of the arguments pitching X-rays against the stethoscope were issues of professionalization and the institutionalization of the new specialism of radiology. In the midst of this development the status of 'seeing' was raised and re-emphasized. This point is well illustrated by one physician writing in 1911: 'When one is equally experienced with the fluorescent screen and the physical methods of examination, it is, I think, easier to detect differences between the two sides (of the thorax) by seeing than it is by hearing.'[11]

Eventually, the X-ray became 'black-boxed' in medicine, its functionality and usefulness completely taken for granted. It came to be seen as a stable apparatus that produced dependable results. Journals stopped writing about 'new' X-ray technology. But the X-ray's gold standard status was threatened in the 1970s by other technologies which provided more detailed information. These included the computerized tomography (CT) scan, which combined a series of X-ray images taken from different angles to create cross-sectional images of parts of the body, and also the echocardiogram, which captured the imagination of physicians through its capacity to produce two-dimensional real-time images of the heart and of the blood flow through its vessels.[12] These technologies created new possibilities for diagnosis, introducing broader ranges of scale, temporality and fine-grained detail. They were part of medicine's 'imaging revolution', and meant that doctors clustered ever more around the 'i-patient' on screens, whether in the radiology room or at the ward desk, as the images were brought to 'the shop floor'.[13]

Many of these technologies are perceived to have excluded patients, creating forms of expertise and authority of which patients are not a part and introducing distance and detachment because the images are assessed away from the bedside. Yet as technology advances in clinical care, new instruments are arguably bringing doctors closer to patients, returning us to our contemporary example of bedside (or point-of-care) ultrasound. Again, this technology is seen as a possible replacement for stethoscopy, as Jos Roelandt, Professor of Cardiology at Erasmus University Rotterdam, writes: 'Future generations of doctors will find it hard to believe that, in 2013, many clinicians were still relying on the vague findings of a 200-year old traditional examination and were compromising clinical efficacy when direct information was available from point-of-care echocardiography.'[14]

The stethoscope here is still pitted as an inferior technology, which has been (or should have been) surpassed by accurate, clear, objective images produced by an exciting (if expensive) new technology. It is striking, however, that it is auscultation that is so often invoked as the retrograde alternative to medical innovation. Its critics inadvertently reveal just how important the technique still is as a baseline in medical practice.

Challenges of Vocabulary

A persistent challenge to the stethoscope over time has been the fact that auscultation depends on a high degree of skill and interpretative ability. Auscultation findings are often regarded as very subjective compared to those gained by the more 'objective' tests detailed above. This is partly due to the complexities of communicating about sounds and a historical lack of consensus on how to name auscultation signs.

In the medical literature, lung and heart sound nomenclature has long been seen as imprecise and ambiguous. For two centuries the names which have been given to sounds have been derived from those coined by Laennec and translated into English by the physician and advocate of auscultation John Forbes. His task was made additionally complex because Laennec adopted the practice of using Latin terms at the bedside lest the French terms be comprehensible and frightening to patients.[15] At the same time, Laennec struggled both to describe some sounds and to establish a standard vocabulary which could be used with the technique of auscultation. Indeed, the complexities of naming sounds threatened to impede the uptake of the practice in the very early stages of its development. Ever since, there have been calls for a more objective naming system that moves beyond what are often seen as fanciful descriptions based on culturally and historically located metaphors and analogies. As late as 1985, for example, an organization called the International Lung Sounds Association attempted to create a universal schema for lung sounds. The terms they produced remain in use, but the system of classifying sounds on which their schema is based is still seen as vague.

Clinicians continue to rely on imprecise ways of sharing knowledge about cardiac and other sounds. Take analogy, for example. Laennec wrote of sounds being '*like* a metallic rattling, *like* a bellows blowing on the fire, *like* a musical tune containing these notes'.[16] In fact, he drew links between stethoscopic sounds and a wide variety of phenomena from the sonic environment, for instance, human and animal voices, natural processes, urban life, hospital practice and music. Lachmund provides a list of examples drawn from the pages of Laennec's treatise:

The voice of *Policinelle*, the ventriloquist; high voices, silvery voices; trembling voices; the voice of a sheep, a voice transmitted through a metal trumpet; the bleating of a goat; the chirp of small birds; the coo of pigeons; the whistle of the wind in the lock of the door; the steady rustle of the sea; the noise of a coach rolling over the pavement; the tinkle of weapons during military exercises; the jingle of a small valve; the crackle of salt being dissolved in a bowl of warm water; the snoring of a sleeping man; the rattle of the dying; the sound produced by a piece of healthy lung-tissue filled with air, which one presses between one's fingers; the crepitation of a dry bladder which is being inflated; the rumble of a drum; the sound produced when a string of a bass is beaten with a finger; the vibration of a metallic string which is rubbed with the tip of a finger.[17]

No matter how scientific Laennec's followers attempted to be, they fell back on analogy to describe sounds of the body. Although he made reference to 'the duration of certain sounds, their continuousness or otherwise' and 'their apparent nearness to or distance from the ear', the famous nineteenth-century physician and specialist in cardiothoracic medicine Austin Flint also noted their strong resemblance to other sounds, 'such as the bleating of the goat, the chirping of birds, etc'.[18] In fact, despite trying to be more systematic than Laennec, Flint concluded that sounds could only be described by analogy.

While they continue to be used in contemporary medical education, the cultural and historical specificity of analogies can be problematic. In the Netherlands, pleuritic rubs are described as sounding like feet crunching in snow, a comparison with which students in cold countries are more familiar than those in warmer climates. In Australia, pleural rubs were once described as sounding like squeaking leather, as trams used leather in their suspension; students, however, are no longer familiar with this sound. In the present day, sounds like coarse crackles are linked to that made by Velcro. So while analogies are helpful in trying to produce the kind of sonic alignment we explored in the previous chapter, the dependence on elaborate or shifting metaphors exemplifies the cognitive and linguistic difficulties that are inherent in making such alignment come about.

Knowing words for a sound is one step, but associating them with sensory experiences is another. Just like perfumers and other sensory experts, doctors need to be trained to match sensations to words and types of sounds they hear through the stethoscope.[19] Nowadays students of auscultation start to learn the general kinds of sounds produced by hearts, lungs and stomachs. They learn about how to elicit sounds and where to find them, for example particular cardiac murmurs at specific points or on particular parts of the chest. They then move on to sensations that are quite obvious or easy to contrast: loud and soft, dull and resonant, long and short and so on. As they go through their training students learn smaller and smaller differences. In auscultation teaching they learn to be affected by distinctions they might not previously have noticed and didn't have the words to describe before: as the cultural theorist Bruno Latour writes, 'the more you learn, the more differences exist.'[20]

Yet the discernment of fine differences presents a challenge to the objectivity that has been sought in medicine since the nineteenth century.[21] Time after time, instruments have been used to replace subjective measurement and judgement in medicine. The thermometer, for example, gives an objective reading of temperature that differs from a purely tactile assessment with the hand. Two thermometers will give the same reading if they are properly calibrated, but how do you calibrate subjective human listeners? The ongoing search for a reliable shared vocabulary is one attempt to achieve this consistency between listeners to ensure that stethoscopic findings can stand up to scrutiny from those seeking more objectivity and clarity. Proponents of the physical examination argue, however, that there is much more to stethoscopic listening than an objective assessment of the patient's condition. It is also closely bound up with elements of presence and humanity.

The Ritual of Auscultation

The technological creep of machines such as X-rays, CT scans and ultrasounds has in large part led to a movement in clinical medicine which advocates a return to the physical examination and which celebrates its endurance. As cries for the death of the stethoscope ring out, there is a counter-cry from the physicians who point to the value of 'traditional' clinical skills. These skills are seen to be lacking for a number of interconnected reasons: over-reliance on the competing technologies detailed above; the limited time that consultants now spend on wards, which reduces opportunities for teaching; loss of expertise in the teaching of the sonic signs of illness; the changing nature of patient hospital stays (which tend to be much shorter than before, meaning that students have less time to see patients on the wards); and the rise in litigation, which leads to more testing. It is also suggested that a reduction in the assessment of students' skills in physical examination in parts of the world such as the United States has led to a decline in their use. As a result, many arguments made in favour of maintaining the physical examination are in fact coming from America.

Abraham Verghese, an Ethiopian-trained, Indian-born physician who now works in the United States, is one such proponent of auscultation and the physical examination more generally. Also a writer and teacher, Verghese has eloquently argued that the physical examination is a ritual that must be kept in medicine. He makes his plea through regular articles in high-profile medical journals. With the help of colleagues, he has set up an initiative called Presence: The Art and Science of Human Connection. This programme aims to foster multidisciplinary collaboration to strengthen the human dimension of medicine. On the organization's website, Verghese is quoted as saying: 'In a world where we are hyper-linked by technology, we are increasingly separated by a lack of human connection. Even as technology is critical to quality and safety in the delivery of care, it inadvertently creates barriers between the patient and the health-care team.'[22] Verghese bemoans the rise of the 'i-patient', describing how he finds that ward rounds often consist of looking at print-outs and images rather than attending to the patient lying nearby. He draws upon his experience in India and Ethiopia to highlight how doctors in these countries often have very inexpensive equipment to make diagnoses: their hands, their ears, their stethoscopes. On his own wards at Stanford, he saw that medical students' and doctors' skills in clinical examination were poor.

Verghese and a group of his registrars who were enthusiastic about the physical examination started an initiative to reinvigorate clinical skills among colleagues and students. They began to conduct regular teaching rounds focusing on the bedside examination. This innovation grew in popularity and soon the 'Stanford 25' was born: a series of 25 physical examinations that the group now teaches through online videos and demonstrations to educators and students around the world. These tutorials give students guidance on how, for instance, to position the patient, handle the stethoscope and find heart sounds. Such resources are working in some ways to stem the threat to the stethoscope associated with the rise of competing technologies.

Although Verghese's voice may be widely heard and his initiatives well-funded, it remains to be seen whether work like his will keep the physical examination, including auscultation, alive in clinical medicine. Perhaps the key to its success may be in the ways in which auscultation engenders a particular kind of presence (the very name of Verghese's initiative) with patients that is less about clinical diagnosis and more about the physicality and intimacy of touch in an era of increasing medical distance and detachment. Ironically, given that Laennec originally felt one of the benefits of his *cylindre* was that it created distance from fleshy, smelly, unhygienic bodies, the stethoscope may now be bringing the doctor back closer to the patient, fostering intimacy.

The movement of Verghese and others towards physical exam-ination skills teaching is encouraging for those who seek the continuity of these arts in medical practice. However, as Professor of Cardiology at the University of Cincinnati Robert J. Adolph observes in his letter to the editor of *Chest*, a pulmonary disease journal: 'Those of us who learned bedside skills in the 1950s and 1960s and developed and maintained these skills in parallel with the growth of high-tech diagnostic procedures clearly represent a dwindling coterie.'[23] While we authors believe there is still a place for the stethoscope in clinical practice, particularly in helping to train doctors' sensory awareness and embed skills in attention that are important in diagnosis, the future of auscultation in hospitals does look bleak. In specialisms such as cardiology and respiratory medicine there is a growing awareness that auscultation could be just a preliminary, cursory technique for determining whether or not abnormalities are present. If something abnormal is heard, it could be investigated further using other technologies. There is certainly far less reliance on the stethoscope in today's hospitals than there was in the past, and correspondingly less need for real expertise in auscultation. While often taught passionately to students, the seduction of echocardiograms and ultrasounds are luring qualified doctors away from stethoscopic practices.

In an opinion piece in the *New England Journal of Medicine*, a group of physicians write that:

> Today ... the tether is fraying, and the auditory stethoscope is all but obsolete. Auscultation is a fading art. Physicians who hear murmurs call for an echocardiogram with little additional effort at sound differentiation – falling prey not only to the loss of physical examination acumen and the allure of images, but also to a belief in the physical exam's futility.[24]

As competing technologies and the rise of artificial intelligence in medicine offer increasingly seductive claims of objectivity, data amalgamation and possibilities for sharing information with patients, fewer doctors will develop their skills in physical examination with the stethoscope, and as such there is a risk that the stethoscope will seem even more redundant to the majority.

In this chapter we have examined claims for the reported demise of the stethoscope and also the counterclaims of those who believe it has never been more important to retain the stethoscope in the doctors' toolkit. What we did not highlight was that many of these positions emerge from hospitals, clinics and medical schools in the United States, Britain, Australia and the Netherlands, where our own fieldwork has been focused. But what about elsewhere in the world? Can we really claim that the stethoscope is dead in places where doctors must make do with whatever instruments they have on hand, where ultrasound scans are not even in the realm of possibility? While there might be truth to the assertion that the stethoscope is dying in some contexts, in others we find it is very much alive.

7

Improvisation

A 'technology creep' poses a challenge to stethoscopic listening in clinical practice in many parts of the world, but in places where medical equipment is unaffordable and difficult to maintain the stethoscope remains prominent. It also sometimes wields a different kind of power and meaning than in European and American hospitals and is used in conjunction with forms of healing that are not biomedical. This chapter looks at stethoscopic practices that are very different to the examples we have used in the book so far. It also highlights the diversity of stethoscopic activities by looking at the ways in which the stethoscope is being reworked into contemporary forms, for instance through digitization, the iStethoscope app or 3D printing. New versions of the stethoscope may hold promise and hope for clearer, more mobile recordings that can be shared among doctors and researchers, experts and novices, keeping stethoscopes relevant in biomedicine. But these technological transformations also have wider consequences for medical practice and raise questions about the balance of distance and intimacy, subjectivity and objectivity, as well as fidelity and standardization in medical work.

Auscultation in the Global South

Despite decades of social science and humanities research about doctors, to date, the vast majority of what is known about medical practice, and especially stethoscopic listening, comes from research in affluent regions in Europe, North America and Australia. Our own histories of the stethoscope in this book attest to the strength of this dominant and prevailing narrative. In such settings the stethoscope is generally being used less and less by doctors. In the Global South, however, the story is different.

We have seen in Chapter Two how the stethoscope was incorporated into Ayurvedic medicine. Anthropologists have also reported how *chota* doctors in Pakistan – usually hospital workers with access to stolen pharmaceuticals who set up private clinics – rely on stethoscopes to lure clientele,[1] while in Malawi healers conducting divination have used dummy stethoscopes along with white coats to shore up their authority.[2] In Papua New Guinea the anthropologist Ian Harper witnessed another form of stethoscope use beyond biomedical practice, and documented this in an ethnographic piece co-written with anthropologist Alice Street.[3] He was spending time with both biomedical doctors and traditional healers and found that rather than these being two separate systems of healing, the boundaries often blurred. The healers wanted to increase their power and influence in their communities by tapping into biomedical knowledge. One way in which to do this was to use the stethoscope. Harper saw how healers would use the stethoscope to 'sweep out spiritual forces' associated with soul-loss, for example. The procedure involved sweeping the stethoscope back and forth across the patient, along with chanting to call back his or her soul. At the same time, the healer was also giving electrolyte therapy, so biomedical and spiritual healing were not incompatible. Harper points out that it was the local doctors who objected to this form of stethoscope use, which they saw as being based purely on superstition.

With tight funding and intense pressure from global health organizations to improve maternal health outcomes, doctors working in obstetrics in the Global South often struggle with differences in the meanings that they and their patients attribute to bodies, symptoms and technologies. The stethoscope, rather than acting as an alienating instrument, can actually be used to bridge these differences. In Mexico, anthropologist Rebecca Howes-Mischel observed how doctors would use the foetal stethoscope to help bring mothers into 'acquaintance' with their foetus.[4] These women felt, like those who went to the healers in Papua New Guinea, somewhat reassured by the use of the stethoscope.

Maternal health is a common site of stethoscope use around the world today, particularly in locations where power supplies are not always constant and machines which provide electronic or digital traces of the baby are not reliable. In South Africa, rural midwives not only find the stethoscope practical, but use the foetal stethoscope to bring about connections between mother and foetus. In his ethnographic work, Gavin Steingo describes how midwives today continue to use the Pinard horn, a short monaural stethoscope made of wood or plastic which sells for only a couple of dollars.[5] He shows that stethoscopes rose in prominence in Africa in the late twentieth century and remain now as a part of the medical toolkit of many healthcare providers. Steingo points out that antelope horns were traditionally used by healers in southern Africa. He suggests that it may be that the use of the foetal stethoscope, which looks and feels similar to such horns, connects to local tradition for carers and their patients. It is clear that the stethoscope is an instrument which practitioners and patients can use in ways that align with their own ideas of care.

In her book on medical education in Malawi, the anthropologist Claire Wendland expands on how, for the medical students she spent time with, the stethoscope and the white coat were also emblematic of biomedical power and glory.[6] Their allure sometimes featured in stories students told her about why they chose to become doctors. At the same time, however, local medical students often felt frustrated by the constraints of their own healthcare system, so that the stethoscope came to represent a kind of medicine that they learned about but didn't always see happen in practice in the hospitals and clinics where they did their training. This frustration was sometimes compounded by encounters with visiting medical students from affluent countries who wielded the stethoscope as if it was an instrument from a bygone era.

Wendland extrapolates on these reflections in an essay on the stethoscope as an object of circulation.[7] In Malawi, stethoscopes circulate, for example, as part of gift exchanges. These are powerful objects which can also be markers of inequality – who owns a stethoscope, and who does not. She found that some Malawian medical students would rather avoid the physical examination classes which involved auscultation than be shamed for not having a stethoscope. Her own stethoscope was a gift from a pharmaceutical company, something that a few of her fellow medical students felt uncomfortable about. Several of her American classmates gave their stethoscopes away during their overseas medical electives to practitioners who had little access to medical imaging technology.[8]

Some doctors from the Global South often take a lot of pride in their auscultation skills. A very experienced Sri Lankan doctor in one Australian hospital told Anna stories about his auscultation abilities during breaks in the more menial tasks he was employed to do as a junior doctor. The stories helped him adjust to what he perceived as his inferior position in the hospital hierarchy. One time he described how he had saved both his son and his wife by checking her blood pressure when she was pregnant, then using his status as a doctor in Sri Lanka to get her the treatment she needed. We also saw in the last chapter that Abraham Verghese, the Ethiopian-trained doctor now at Stanford University, is one of the biggest advocates of the physical examination in the United States. Dr Felix Konotey-Ahulu, now a professor of Human Genetics at the University of Cape Coast in Ghana and a consultant at the Harley Street Clinic, London, also offered a passionate response to criticism of the relevance of physical examination in contemporary clinical practice. He describes astounding his American colleagues many times over with his skills in diagnosis based on stethoscopic examination, publishing on novel bruits associated for example with Parkinson's disease. He writes, 'the stethoscope is there for life and will remain for good.'[9]

Hacking the Stethoscope

In some places at particular moments even the humble stethoscope is a precious and rare commodity. During the fallout of the Israeli invasion of Gaza in 2014, the Canadian doctor Tarek Loubani was working in extremely difficult conditions as an emergency room doctor. Blockades on imports across the Israeli and Egyptian borders had contributed to a severe lack of vital supplies in Gaza. Left to care for many badly injured people with only two stethoscopes in the emergency room, Dr Loubani had to hold his ear to the chest of patients. He found this unacceptable and began to consider ways in which to make technologies more affordable in war-torn parts of the world. Impressed by the quality of his nephew's plastic toy stethoscope back in Canada, Loubani set out to rework the stethoscope using the technology of 3D printing.[10]

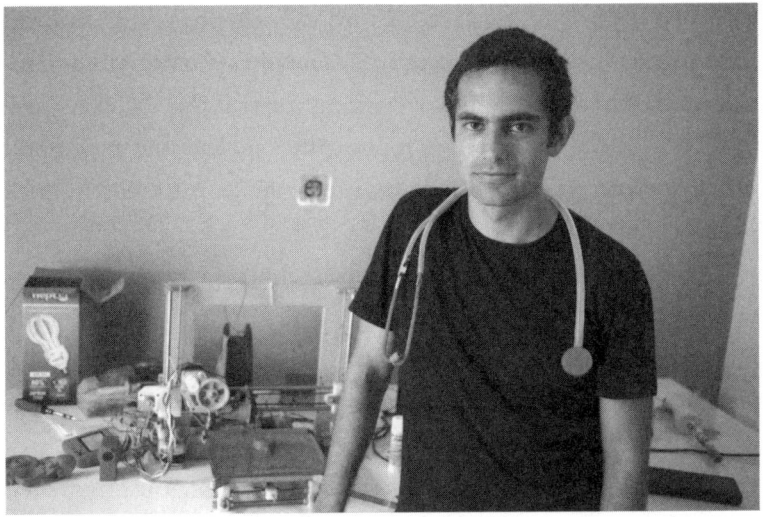

Dr Tarek Loubani with his 3D-printed stethoscope, 2015, Gaza City.

The 3D-printed stethoscope he produced consists of a head, ear tubes and ear tips all made in plastic. It costs CAD$5 to make, though Loubani does not want the monopoly on production and has put the plans for the stethoscope online so that anyone with a 3D printer can make them. A team of surgeons and technicians have conducted audio tests and found that the device offers superior quality compared to other popular models of stethoscope such as the widely used Littmann Cardiology 3. Canadian doctors have also attested to its effectiveness and fidelity. The stethoscope is now being manufactured and given to many doctors in Gaza who have previously not had access to stethoscopes. Once it has met the required approvals, the plan is to roll out this type of stethoscope to medically under-served regions globally.

Loubani has no intention of stopping at the stethoscope. His overarching goal is to apply the principles of open-source software development to out-of-patent medical devices. He has already been working on other equipment, from a loom for weaving gauze to an otoscope for inspecting ears, all with the aim of making medical tools more affordable and accessible. Loubani's work might be considered part of a wider trend of hacking and making. 'Hackers' and 'makers' – those who attempt to move beyond the confines of pre-manufactured products – are increasingly appearing in institutions like hospitals through organizations such as Maker Health.

Maker Health started when some technological entrepreneurs noticed the tinkering work that was going on in hospitals in the Caribbean, where doctors and nurses would repair or make medical equipment using simple solutions. One of the solutions they witnessed was the replacement of a diaphragm of a stethoscope with a round segment of overhead slide projector material. The group then realized that the same kind of tinkering was happening in hospitals in the United States, usually those with fewer resources and less centralized control. The group is currently working to establish 'maker spaces' in hospitals. It has various posts on their website for how to 'hack the stethoscope'. For example, they show how to make replacement earpieces out of Sugru, a mouldable glue that turns into rubber, and how to make a belt out of rock-climbing carabiners to carry a stethoscope.

The stethoscope is an invention of makers, and continues to be revised according to the needs of its users. But it is not only doctors who are 'hacking' the stethoscope. A farmer by the name of Mark Himes had a buried water pipe which sprung a leak.[11] He avoided the expensive job of digging up the entire pipe by attaching a stethoscope to the bottom of a paper cup and attaching the cup to a 1.5-metre-long piece of PVC pipe. He used this simple listening device to find the affected section of pipe through metres of solid earth.

We also see the stethoscope being hacked for artistic purposes. Graeme Croft is an artist who lives and works in Melbourne, Australia. He has turned his own body into a musical instrument

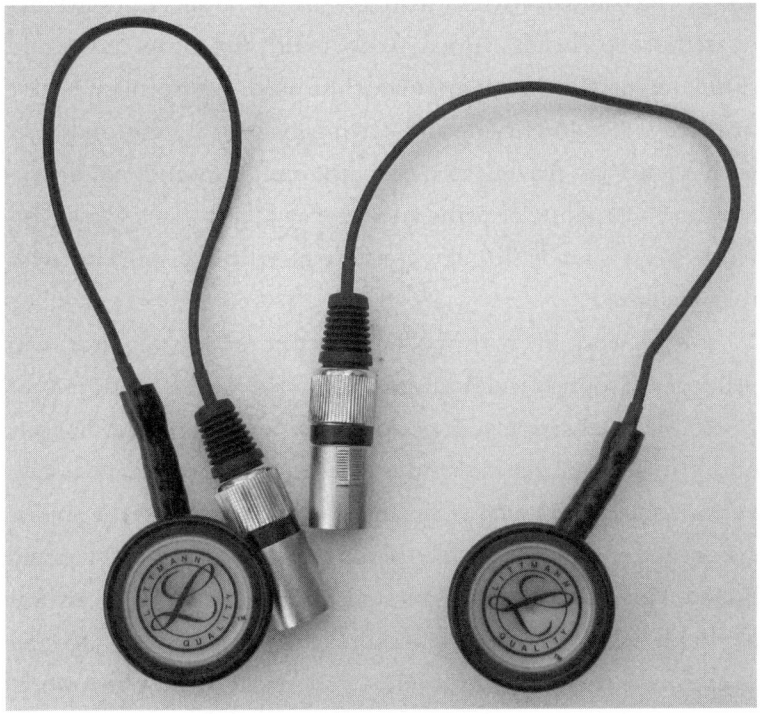

Images showing Graeme Croft's process of constructing the stethoscope-microphone.

using tinkered stethoscopes. Fascinated by both the body's sounds and the role of the recordist's body when recording sounds in the environment, his performances blur the distinction between inside and outside. Croft is both a woodwind player and a doctor, though his specialism is psychiatry. While he may not do a great deal of stethoscopic work, his profession nonetheless made him familiar with the instrument and galvanized him to play with it in his installations and performances.

Croft himself initially tried using commercial electronic or digital stethoscopes, but he found there was too much distortion and limited space for recordings, so he had to make his own device. He worked out how to adjust a standard stethoscope by watching an online video and devised a method of 'fitting lavalier microphones to stethoscope heads', which he used with digital recorders and computer interfaces to manipulate the sound.[12] Performances have involved Croft amplifying and playing his own body sounds live, with a musician playing alongside him on a conventional instrument such as a flute or using an everyday object like a wine glass. Croft 'plays' sounds through gestures and through working with the computer.

Croft's work fits within a longer tradition of using heart and other body sounds (such as those from the inside of vaginas or gastrointestinal systems) in performative musical pieces. Pioneers of electro-acoustic music have used the sounds of heartbeats, as have performance artists such as Stelarc, who is interested in the artistic cyborg and creates pieces which blur boundaries of technologies and bodies. He has used, for instance, recordings of his own heartbeat alongside other sounds, such as those of his own stomach, in order to explore extensions of the body – a recurrent theme in his work.

Electrification and Digitization

The fact that the stethoscope has been constantly reworked is representative of the 'tinkering' or 'doctoring' that goes into medical care more generally. Medical practitioners have always adapted and reworked tools in order to make them more efficient at particular tasks. The history of the stethoscope, beginning with Laennec's rolled paper experiments, is testament to this. We saw how the stethoscope evolved from a one-eared listening device into a binaural version. We now look at how the stethoscope has been further reworked through electrification and digitization in recent medical history and in contemporary life.

Interest in electrification of the stethoscope emerged in the late nineteenth century, around the time that other sound reproduction technologies such as the telephone, microphone and phonograph were also emerging.[13] Van Drie ties the development of the electronic stethoscope to telephone research where collective listening was the ambition.[14] As mentioned in Chapter Five, in u.s. medical schools in the 1920s collective electronic stethoscope listening systems were installed in lecture theatres, but this complicated system soon morphed into a more portable version that could be taken to the bedside where much medical teaching occurred. By the mid-1930s it had evolved into a model designed to be used by the individual.[15] An article in *The Lancet* described a model of the electronic stethoscope from 1936 made by Samuel Mazza as at once better than the standard stethoscope (for instance, for hearing-impaired doctors, for auscultation of patients with polio in cabinet respirators, for the auscultation of foetal heartbeats and of joints), and also as an instrument through which sounds similar to those heard using a conventional stethoscope could be produced. It seems it was important to pitch the electronic stethoscope as 'better than' but also 'true to' its predecessor. By the 1950s the technology had progressed even further, with better sound quality and the capacity to record sounds offering new possibilities in medical practice and education.[16]

The tinkering of stethoscopes with electrical technology has allowed sounds to travel. Telemedicine is a growing field in many parts of the world, particularly those with rural and remote communities, for instance North America, Australia and Scandinavia. In Norway, a group of researchers studied the usefulness and accuracy of heart murmurs that were recorded and then emailed as digital files for remote assessment. The study, based at University Hospital in Tromsø, was founded on the observation that doctors in small rural communities were often inexperienced in assessing heart murmurs, and thus needed to rely on remote help.[17] Sounds of patients in the hospital were recorded using a sensor-based electronic stethoscope. Qualities such as amplification were fixed, and the sounds were sent to four paediatric cardiologists across Norway to

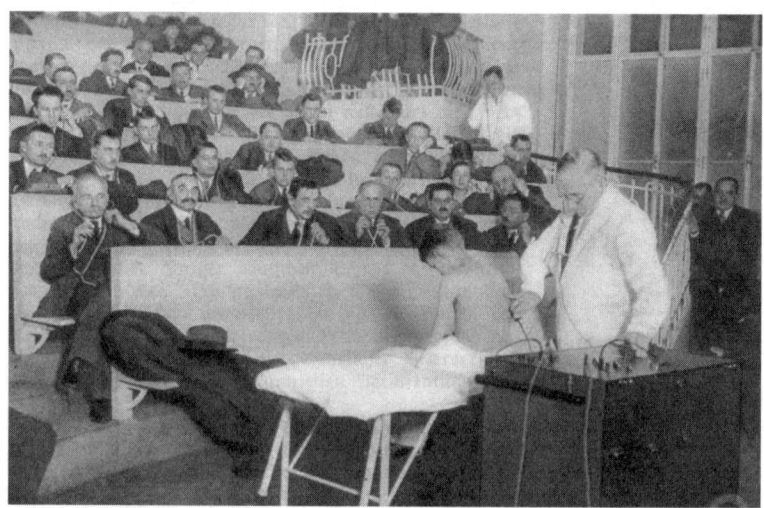

Karel Frederik Wenckebach demonstrates a boy's heartbeat to medical students while they listen through stethoscopes connected to a machine, c. 1925.

listen to. The researchers found that the recordings offered good sensitivity and specificity, and that inter-observer and intra-observer variability were low (that is, there was little difference within groups of observers and between groups of observers). Such studies add legitimacy to the use of telemedicine referrals for patients with heart murmurs. These are argued to be safe and to save time. They also allow us to see stethoscopic listening as something that is increasingly distributed or dispersed, with listeners separated by vast differences but remaining linked through their attention to a sound. Telemedicine has become more common globally in recent times because of the COVID-19 pandemic.

The desire to augment the humble acoustic stethoscope with electrical charge so as to enable amplified sounds of the body to be transmitted beyond the individual listener continues nowadays in the form of the digital stethoscope. As with the electrical version, digital stethoscopes have also been claimed to make clearer, more mobile recordings that can be shared among doctors and researchers, experts and novices. But digital stethoscopes are also promoted for features such as visual displays, playback features, data storage and database building capabilities.[18] They have been on the market since at least the early 2000s, an early example being the eSteth, designed in France by a company called IRIS and acquired by an American company in late 2000.[19]

At the time of writing, digital stethoscopes are still not common in hospitals. Major stethoscope companies such as Littmann offer digital stethoscopes to doctors for free trial periods, so that they can get used to and experiment with its features. These trials highlight the extra work required to sell this product to a sceptical body of customers, but also the process of reskilling that is required when using the instrument. Potential users have to try it out (and not just once) to learn its potential, because the digital stethoscope differs in subtle ways from the non-digital stethoscope with which most doctors are more familiar. The digital stethoscope looks in many ways like its analogue counterpart, but it can also amplify sound beyond the 'normal' level and record sounds which can be played back on the computer via Bluetooth technology. Many doctors Anna spoke to during her fieldwork are wary of the digital stethoscope. Some feel superstitious that they would not be able to diagnose properly without their more familiar versions. Others say the better sound quality and noise-cancellation in fact make it more difficult to diagnose, as the sounds are clearer than they are used to and thus difficult to correlate with what they already know and have been trained to hear.

If doctors are sceptical as to the benefits of the digital stethoscope, they are even more cautious in relation to technological developments such as the iStethoscope app. Introduced in 2010, at the time of writing this app has been downloaded more than three million times. In principle at least, it allows users to record and listen to their own body sounds by placing the microphone in their smartphone at key auscultation points. As with other digital stethoscopes, recordings can be made and then replayed. The professional version of the app 'app'arently has other features, for instance, it enables the user to see a spectrogram of the recorded audio and send the sound over email in different formats. The description emphasizes, however, that the app 'is intended to be used for entertainment purposes . . . and should not be used as a medical device'. User reviews also indicate that getting clear recordings is particularly difficult and overall satisfaction with the gadget is evidently low. It is interesting to note that the iStethoscope app description clearly states that the difficulties of learning to auscultate with a smartphone parallel some of those that have been involved in learning auscultation since the outset and which have been discussed by doctors and medical students for more than two centuries:

WARNING! It takes time to learn how to listen to your heart. You may need to watch the youtube videos. Don't give up! If you are patient, this app will help you to learn the necessary skills. It takes physicians years to learn how to use stethoscopes correctly and the iphone microphone is smaller than a conventional stethoscope diaphragm. You cannot place the microphone randomly and expect to hear your heart – you must place it very accurately. To hear your heart you must learn where to listen![20]

Yet smartphones are increasingly becoming part of the doctors' toolkit, as well as a technology regularly used by patients to monitor and record their own health. It may well be that more user-friendly and reliable auscultatory apps will appear in the future, generating new iterations of the stethoscope. Since 2020 we have also seen digital stethoscopes with integrated artificial intelligence diagnostic support entering the marketplace. Research shows high sensitivity and specificity for detection of heart murmurs.[21] Digital stethoscopes in combination with machine learning technologies also promise to generate powerful new stethoscopic trajectories.

The Pandemic

This book was written largely before the COVID-19 pandemic. So how has this extraordinary event, which catapulted medicine well and truly into the digital age through online medical education, telemedicine consults and the explosion of apps and digital tools for healthcare, impacted the life, death and reinvention of the stethoscope?

COVID-19, the novel coronavirus with which the world is now all too familiar, is a disease of the respiratory tract. From the very early stages of the pandemic there was concern among healthcare workers and in medical journals about the use of the stethoscope in examining patients suspected to have the disease, due to the risk of infection of the healthcare worker and cross-contamination between patients.[22] Stethoscope contamination with microbes including antibiotic-resistant organisms and germs which cause gastrointestinal diseases, such as E. coli, is widely reported in medical literature.[23] Research also shows that clinician's adherence to recommended protocols for cleaning stethoscopes is low.[24] Some doctors highlighted that the value of handheld ultrasound was ever more salient in the pandemic as an alternative to the stethoscope. These clinicians suggested that the ultrasound can be covered with plastic sheets, and it may be that the wireless nature of ultrasound is easier to use in the context of personal protective equipment (PPE).[25] Others, however, were not so quick to dismiss the stethoscope altogether.

Pointing out that it is no surprise that the fate of the stethoscope is once again being called into question during the COVID-19 pandemic, American and Japanese physicians conducted an overview of its contemporary relevance. They argued that the infectious disease reasons for abandoning the stethoscope cannot completely override the importance of the tactility that comes with the act of auscultation. PPE, after all, was felt by some patients to present very serious barriers to interpersonal communication, and some of the most distressing stories from the pandemic were those of patients dying alone, surrounded by machines (or not, depending on where they died), separated from the physical touch of carers due to the contagious nature of their disease. Healthcare workers, the members of society who were witnessing this pain at first hand, sought simple ways to try and alleviate their patients' sense of isolation. They wheeled them to see views outside the hospital buildings and filled rubber gloves with warm water to simulate the touch of a human being. The pandemic may have highlighted the infectious disease risk of technologies such as the stethoscope, but it also raised the importance of proximity and touch for our sense of being cared for.

Improvisation, hacking and tinkering may well ultimately provide a solution to the hygiene problems created by the stethoscope, freeing up opportunities for contact. For instance, researchers are looking at how parts of the stethoscope such as the diaphragm could become single-use items (though this adds stethoscope components to the growing list of disposable clinical equipment used in hospitals). It is also interesting to note that the stethoscope continued to be used throughout the pandemic as a symbol of medical advice, healthcare services and public health in general. As millions of people became ill, the stethoscope, in this symbolic sense at least, was shown to be in good health.

Conclusion

In this book we have shown that the stethoscope has been a key instrument in the development of biomedicine, as well as in other healthcare practices, and that it continues to be an important presence. In fact, we have argued that the stethoscope remains central to core concerns of medicine today. It raises questions about what kind of interactions we desire between doctor and patient. While originally perceived to facilitate the creation of distance and detachment (in addition to its role in facilitating the detection of anatomical change through sound), the stethoscope is now linked to notions of closeness and connection. Some doctors and nurses cling to their stethoscopes as symbols of a more humane medicine which takes time to get to know the patient through history-taking and physical examination. Ironically, the stethoscope, first seen as a tool which silenced the patient, now implies that a doctor is there to listen. Increasingly, diagnoses are made at the computer, and advances in artificial intelligence mean they may soon be made *by* the computer. We believe that the ongoing debate around the life of the stethoscope tells us a lot about the wider concerns that exist about the directions medicine is taking today.

We have seen in this book that the stethoscope is not only something owned by doctors. Other healthcare professionals use stethoscopes: nurses, midwives, physiotherapists and speech therapists to name a few. People who have not been trained in medicine use, tinker with and adapt stethoscopes too: engineers, farmers, soldiers and artists, for example. While on the one hand the stethoscope denotes a doctor's expertise and skill, on the other hand the technique is also owned by outsiders to the medical profession, just as patients are increasingly recognized as experts in their condition and citizens as scientists.

The stethoscope is considered by some contemporary doctors to be an antiquated instrument, a relic of a bygone era. But the tool has undergone, and continues to undergo, surprising transformations that give it an ongoing presence around the world. Reworked versions show how the stethoscope lends itself to continual experiment and reinvention. Throughout this book we have tried to suggest that stethoscopes live on not only through tinkering and hacking practices of medical professionals, but in the minds of film-makers,

A client consults Frank, a spirit medium, in Ben Steiner's short film *The Stomach* (2014). The stethoscope is used to listen to voices issuing from inside Frank's body.

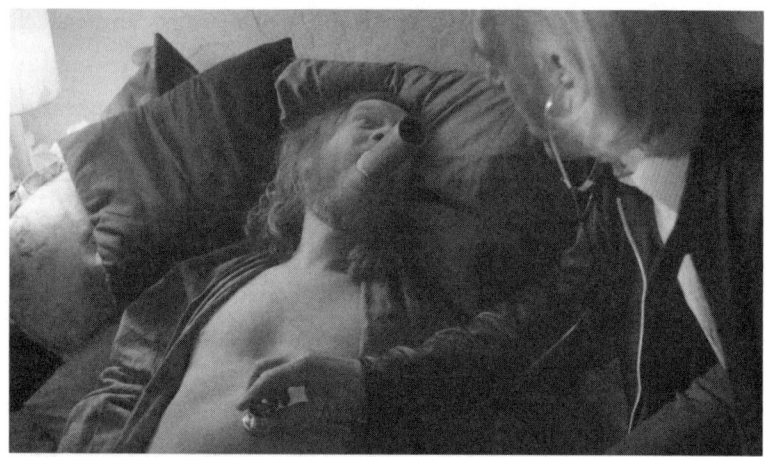

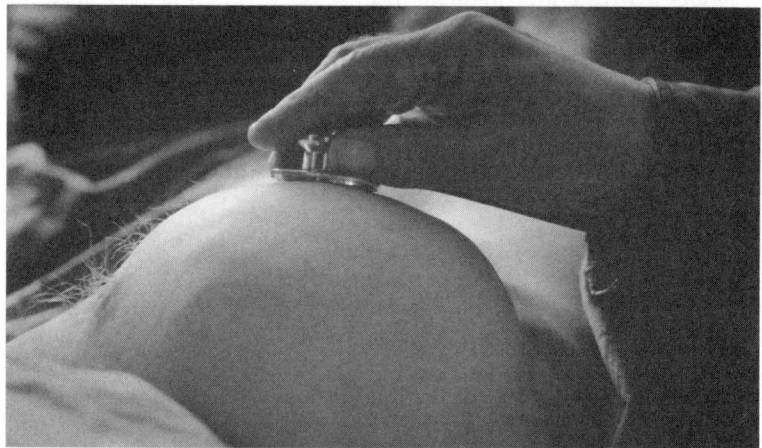

The Stomach (2014).

novelists, poets and artists. We have been keen to show how the stethoscope is a creative resource and has provided inspiration to people working in different modes, from the metaphorical to the interactive, as well as in a range of moods, from the playful, comic and irreverent to the critical, reflective and sombre. The instrument continually reappears as a portal to imaginative depths and expanses, a tube into the hidden visceral depths of bodies and minds.

Just as it has been reworked in a wide variety of clinical and practical directions, then, the stethoscope has also been taken in a variety of artistic ones. Indeed, it has not been possible to provide an exhaustive list of its artistic uses in this book. For instance, the instrument is also evoked in songs by The Adverts, Robyn Hitchcock, Peter Gabriel and Kiki Dee. It features extensively in visual culture and is used to enhance characterization and action in films and numerous medical TV series, not to mention the Internet phenomenon of medical and, specifically, 'stethoscope porn'. Its association with the 'heart', with depth, with the realm of the invisible, mysterious and unknown (or partially known), but also with the physical and corporeal, make the stethoscope hugely versatile and good to think with, good to carry with us.

There will always, we believe, be some form or iteration of the stethoscope. Whether it will look similar to the version now donned by actors and featured in medical photo shoots remains to be seen. Inventors and tinkerers will continue to rethink the stethoscope in unexpected ways. A recent Kickstarter project called the Stemoscope, for instance, shows the power of stethoscopic listening as a practice beyond medicine. The Stemoscope is a round disc which is attached via Bluetooth to a smartphone, allowing the user to listen, without tubes, to the sounds of the heart or lungs, and to amplify, record and share these sounds via other technologies. Importantly it is not only hearts that the user can listen to, but animals and trees, too. The Stemoscope is a way of listening in on the world. Maybe, the designers postulate, this is a way for humans to become more in tune with their surroundings and the non-human inhabitants with which they share them.

The originators of the Stemoscope touch on an important point here, namely that the stethoscope is an attitude as much as an object. A stethoscopic act is one characterized by careful listening

Stethoscopes are common props in television shows, which are known to be highly influential in the lives of many medical students. Some identify closely with the characters. Pictured here are the television series *Scrubs* ('My Growing Pains', 2007) and *ER* (1994).

and attention to what lies beneath the surface. This is why we see auscultation appealing to poets and artists, as well as doctors and engineers. It is why it has become a way of understanding, for instance, changes in the world due to the pandemic such as lock-downs. As anthropologist Shannon Mattern writes in an essay on urban auscultation, COVID-19 was a disease that reshaped our sound-scapes with its unwanted coughs and sneezes, the ventilators in hospitals, applause for healthcare workers and quiet streets. Mattern muses, 'as physicians monitor the rattle of afflicted lungs, the rest of us listen for acoustic cues that our city is convalescing, that we've turned inward to prevent transmission.'[1] She goes on to write that an auscultative attitude may be central to how we navigate these new urban landscapes:

> In other words, *how* we listen to the city is as important as what we are listening for. Amid the rise of artificially intelligent, algorithmically attuned ears, scoring the city in accordance with their own computational logics, we humans need to better understand our own acoustic agency so that we can make thoughtful choices about how to supplement our ears with machinic ones . . . A polyphonic mode of distributed listening helps us appreciate how our actions – making music and noise, building and maintaining infrastructure, tracking and monitoring fellow citizens, creating acoustic space for bodies to rest and heal – reverberate across time and space, and beyond the range of human ears.[2]

The literary theorist Ben De Bruyn writes that stethoscopic perception 'permits the attentive individual to access the invisible lives of others', making writing a technology of amplification in itself.[3] He suggests that George Eliot's famous quote about what it would feel like to perceive the world with an extra-human range of faculties – 'It would be like hearing the grass grow and the squirrel's heart beat, and we should die of that roar which lies on the other side of silence' – indicates that stethoscopic attention should not be limited to the human, either in reality or in fiction.[4] Thus, whatever physical form stethoscopes might take in the future, stethoscopic listening seems to remain a captivating and enduring mode of attention. Perhaps it is especially valuable today as a metaphor for and a realization of attentiveness in a world filled with distraction?

We leave you with one last image from our fieldwork. In the hospital in Melbourne where Anna did her research there was a box of found stethoscopes. It sat on a lesser-used bench in the busy intensive care unit, a knotted mess of coloured plastic tubes. New stethoscopes were added to it from time to time as a doctor, nurse or physiotherapist laid down their auscultative instrument at a bedside and walked away, distracted by a sideways conversation, a beeping pager or a life-threatening alarm. The box is a little collection of stethoscopes, from cheap ones to expensive graduation gifts. Each gestures towards its own story of heartbeats, murmurs, crackles and wheezes, and of silences. The box is a miniature archive of an instrument which can take so many forms, and has been used to listen in so many ways that we struggle to contain it to just one thing, one moment of invention, one field of practice or one kind of imaginative possibility.

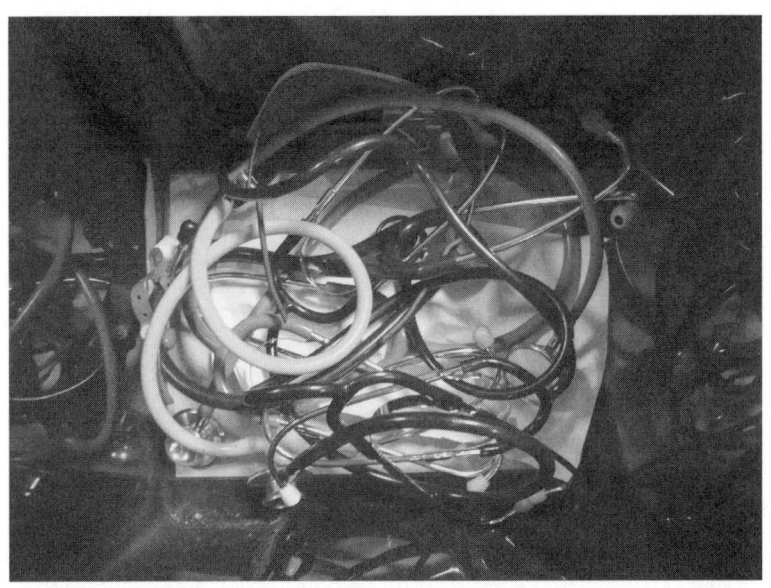

Found stethoscopes in the ICU department of a Melbourne hospital.

REFERENCES

Introduction

1 Jonathan Sterne, *The Audible Past: Cultural Origins of Sound Reproduction* (Durham, NC, and London, 2003).
2 Tom Rice, *Hearing the Hospital: Sound, Listening, Knowledge and Experience* (Canon Pyon, 2013).
3 Anna Harris, 'Listening-Touch, Affect and the Crafting of Medical Bodies through Percussion', *Body and Society*, XXII/1 (2016), pp. 31–61.

1: Revelation

1 Jacalyn Duffin, *To See with a Better Eye: A Life of R.T.H. Laennec* (Princeton, NJ, 1998), p. 18.
2 Ibid., pp. 18–23.
3 Ibid., p. 27.
4 J. F. Halls Dally, 'Life and Times of Jean Nicolas Corvisart (1755–1821)', *Proceedings of the Royal Society of Medicine* (Section of the History of Medicine)', XXXIV/5 (1941), p. 243.
5 Roy Porter, *The Greatest Benefit to Mankind: A Medical History of Humanity from Antiquity to the Present* (London, 1997), p. 256. See also Axel Volmar, 'Sonic Facts for Sound Arguments: Medicine, Experimental Physiology, and the Auditory Construction of Knowledge in the 19th Century', *Journal of Sonic Studies*, IV/1 (2013); and Saul Jarcho, 'Auenbrügger, Laennec and John Keats: Some Notes on the Early History of Percussion and Auscultation', *Medical History*, V/2 (1961), pp. 167–72.
6 Duffin, *To See with a Better Eye*, p. 33.

7 Ibid., p. 117.

8 Ibid., pp. 124–5. See also Albert Mudry and Anders Tjellström, 'Historical Background of Bone Conduction Hearing Devices and Hearing Aids', *Advances in Oto-Rhino-Laryngology*, LXXI/1–9 (2011), p. 2.

9 René-Théophile-Hyacinthe Laennec, *A Treatise on the Diseases of the Chest and on Mediate Auscultation* [1829], trans. John Forbes, 3rd edn (London, 2015), p. 4.

10 Duffin, *To See with a Better Eye*, p. 123.

11 Laennec, *A Treatise on the Diseases of the Chest and on Mediate Auscultation*, pp. 4–5.

12 Duffin, *To See with a Better Eye*, p. 121.

13 Ibid., p. 125.

14 Laennec, *A Treatise on the Diseases of the Chest and on Mediate Auscultation*, p. 37.

15 Ibid., p. 41.

16 Ibid.

17 Duffin, *To See with a Better Eye*, p. 192.

18 Laennec, *A Treatise on the Diseases of the Chest and on Mediate Auscultation*, pp. 5–6.

19 Ibid., p. 6.

20 Ibid.

21 Ibid., pp. 6–7. See also Duffin, *To See with a Better Eye*, p. 129.

22 Laennec, *A Treatise on the Diseases of the Chest and on Mediate Auscultation*, p. 4.

23 Ibid.

24 Ibid., p. 2.

25 Ibid.

26 Ibid., p. 8.

27 Duffin, *To See with a Better Eye*, p. 282.

28 Laennec to Mériadec, 23 June 1826, ADLA MS Cheguillaume no. 23; transcribed in Pia Gourdon Roux de Reilhac, 'La Correspondance de Laennec de 1822 à 1826' (1980), quoted ibid.

29 Michel Foucault, *The Birth of the Clinic: An Archaeology of Medical Perception*, trans. Alan M. Sheridan-Smith (New York, 1973).

30 Ibid., p. 202.

31 Ibid.

32 Ibid.

33 Jonathan Sterne, 'Mediate Auscultation, the Stethoscope, and the "Autopsy of the Living": Medicine's Acoustic Culture', *Journal of Medical Humanities*, XXII/2 (2001), pp. 115–36.

34 Duffin, *To See with a Better Eye*, p. 25.
35 Foucault, *The Birth of the Clinic*, p. 202.
36 Ibid.
37 Duffin, *To See with a Better Eye*, pp. 302–3.

2: Rise

1 Jonathan Sterne, 'Mediate Auscultation, the Stethoscope, and the "Autopsy of the Living": Medicine's Acoustic Culture', *Journal of Medical Humanities*, XXII/2 (2001), pp. 115–36.
2 Projit Bihari Mukharji, 'Akarnan: The Stethoscope and Making of Modern Ayurveda, Bengal, *c.* 1894–1952', *Technology and Culture*, LX/4 (2019), pp. 953–78.
3 Ibid., p. 966.
4. P. J. Bishop, 'Evolution of the Stethoscope', *Journal of the Royal Society of Medicine*, LXXIII (1980), p. 449.
5 Jacalyn Duffin, *To See with a Better Eye: A Life of R.T.H. Laennec* (Princeton, NJ, 1998), p. 213.
6 Roy Porter, *The Greatest Benefit to Mankind: A Medical History of Humanity from Antiquity to the Present* (London, 1997), p. 314.
7 P. J. Bishop, 'Reception of the Stethoscope and Laennec's Book', *Thorax*, XXXVI (1981), p. 489.
8 Peter R. Fleming, *A Short History of Cardiology* (Amsterdam, 1997), p. 90.
9 Ibid.
10 Lester S. King, 'Auscultation in England, 1821–1837', *Bulletin of the History of Medicine*, XXXIII/5 (1959), pp. 446–53.
11 Malcolm Nicolson, 'The Introduction of Percussion and Stethoscopy to Early Nineteenth-Century Edinburgh', in *Medicine and the Five Senses*, ed. William F. Bynum and Roy Porter (Cambridge, 1993), p. 148.
12 Ibid., pp. 150–51.
13 Ibid., p. 142.
14 Ibid., p. 147.
15 Ibid., p. 148.
16 Ibid.
17 Bishop, 'Reception of the Stethoscope and Laennec's Book', p. 490.
18 Fleming, *A Short History of Cardiology*, pp. 90–91.
19 Edward I. Bluth, 'James Hope and the Acceptance of Auscultation', *Journal of the History of Medicine and Allied Sciences*, XXII/2 (1970), p. 203.

20 Nicolson, 'The Introduction of Percussion and Stethoscopy to Early Nineteenth-Century Edinburgh', p. 150.

21 King, 'Auscultation in England, 1821–1837', p. 450. King is drawing here on Thomas Addison, 'Observations of the Diagnosis of Pneumonia', *Guy's Hospital Reports*, II (1837), pp. 57–67.

22 Bishop, 'Reception of the Stethoscope and Laennec's Book', p. 490.

23 Ibid.

24 Ibid.

25 Bluth, 'James Hope and the Acceptance of Auscultation', p. 202.

26 George Eliot, *Middlemarch* [1871–2] (London, 1994). See also 'A Treatise on the Use of the Stethoscope', www.bl.uk, accessed 28 August 2018.

27 Nicolson, 'The Introduction of Percussion and Stethoscopy to Early Nineteenth-Century Edinburgh', p. 150.

28 Ibid., p. 151.

29 Ibid.

30 King, 'Auscultation in England, 1821–1837', p. 448.

31 Porter, *The Greatest Benefit to Mankind*, p. 315.

32 Jens Lachmund, 'Between Scrutiny and Treatment: Physical Diagnosis and the Restructuring of 19th Century Medical Practice', *Sociology of Health and Illness*, XX/6 (1998), p. 783.

33 Ibid.

34 Porter, *The Greatest Benefit to Mankind*, p. 315.

35 Jens Lachmund, 'Making Sense of Sound: Auscultation and Lung Sound Codification in Nineteenth-Century French and German Medicine', *Science, Technology, and Human Values*, XXIV/4 (1999), p. 441.

36 Ibid., p. 431.

37 Ibid., p. 433.

38 Ibid., p. 441.

39 Mukharji, 'Akarnan', p. 966.

40 Ibid.

41 Ibid.

42 Ibid., p. 972.

43 Shigehisa Kuriyama, *The Expressiveness of the Body and the Divergence of Greek and Chinese Medicine* (London, 2002).

44 Mukharji, 'Akarnan', p. 974.

45 Richard A. Reinhart, 'The Stethoscope in 19th-Century American Practice: Ideas, Rhetoric, and Eventual Adoption', *Canadian Bulletin of Medical History*, XXXVII/1 (2020), pp. 50, 76. See also Ibrahim R. Hanna and Mark E. Silverman, 'A History of Cardiac Auscultation

and Some of Its Contributors', *American Journal of Cardiology*, XC/1 (2020), p. 264.

46 Reinhart, 'The Stethoscope in 19th-Century American Practice', pp. 59–63.

47 Ibid., p. 69.

48 Ibid., p. 80.

49 William Osler, 'On the Educational Value of the Medical Society', in *Aequanimitas: With Other Addresses to Medical Students, Nurses and Practitioners of Medicine*, ed. William Osler (Philadelphia, PA, 1905), p. 359.

50 Bishop, 'Evolution of the Stethoscope', p. 451.

51 Ibid., p. 452.

52 Ibid.

53 Arthur Conan Doyle, *Round the Red Lamp and Other Medical Writings,* ed. Robert Darby (Kansas City, MO, 2007), p. 131.

54 Arthur Conan Doyle, *The Adventures of Sherlock Holmes* (Fairfield, IA, 2007), p. 10.

55 Stanley Joel Reiser, *Medicine and the Reign of Technology* (Cambridge, 1978), p. 37.

56 N. P. Comins, 'New Stethoscope', *London Medical Gazette*, IV (1829), p. 427. Quoted in Nicolson, 'The Introduction of Percussion and Stethoscopy to Early Nineteenth-Century Edinburgh', p. 147.

57 Ibid., p. 146.

58 Bishop, 'Evolution of the Stethoscope', p. 453.

59 Ibid.

60 Ibid.

61 Mukharji, 'Akarnan', p. 974.

3: Reach

1 Jacques-Alexandre Le Jumeau de Kergaradec, *Mélanges de médecine*, XXVI/3 (1822). Quoted in J.H.M. Pinkerton, 'Kergaradec, Friend of Laennec and Pioneer of Foetal Auscultation', *Proceedings of the Royal Society of Medicine*, LXII/5 (1969), p. 482.

2 Ibid., p. 483.

3 J.H.M. Pinkerton, 'John Creery Ferguson: Friend of William Stokes and Pioneer of Auscultation of the Fetal Heart in the British Isles', *British Journal of Obstetrics and Gynaecology*, LXXXVII/4 (1980), pp. 257–60.

4 Julie Harrison, 'Auscultation: The Art of Listening', *Midwives Magazine* (February 2004), online at www.rcm.org.uk.

5 Anon., 'Review of *A Memoir on the Diagnostic Signs Afforded by the Use of the Stethoscope in Fractures and in Some Other Surgical Diseases*. Translated from the French of Professor Lisfranc, with Notes and Additions by J. R. Alcock. London, 1827, pp. 46, 18mo. With a Plate', *American Journal of the Medical Sciences*, II (Philadelphia, PA, 1828), p. 188.

6 A. E. Garrod, 'On Auscultation of Joints', *Journal of the Royal Society of Medicine*, IV: Medical Section (1911), p. 35.

7 René-Théophile-Hyacinthe Laennec, *A Treatise on the Diseases of the Chest and on Mediate Auscultation*, trans. John Forbes, 3rd edn (London, 1827), p. 719.

8 Joseph Škoda, *Auscultation and Percussion*, trans. W. O. Markham, 4th edn (Philadelphia, PA, 1854), p. 380.

9 Laennec, *A Treatise on the Diseases of the Chest and on Mediate Auscultation*, pp. 719–20.

10 J. D. Fisher, 'Observations on a Cephalic Bellows Sound', *Medical Magazine* (Boston, MA), II (1833), pp. 144–52.

11 Ian Mackenzie, 'The Intracranial Bruit', *Brain*, LXXXIII/3 (1955), pp. 355–6.

12 Dannie Abse, 'The Stethoscope', www.utmedhumanities. wordpress.com, accessed 28 August 2018. The poem also appears in Dannie Abse, *Be Seated, Thou: Poems, 1989–98* (New York, 2000).

13 Michaela Reid, 'Sir James Reid, Bt: Royal Apothecary', *Journal of the Royal Society of Medicine*, XCIV/4 (2001), pp. 194–5.

14 Stanley Joel Reiser, 'The Science of Diagnosis: Diagnostic Technology', in *The Companion Encyclopedia of the History of Medicine*, ed. W. F. Bynum and R. Porter (New York, 1997), p. 832.

15 Jens Lachmund, 'Between Scrutiny and Treatment: Physical Diagnosis and the Restructuring of 19th Century Medical Practice', *Sociology of Health and Illness*, XX/6 (1998), p. 788.

16 P. J. Bishop, 'Reception of the Stethoscope and Laennec's Book', *Thorax*, XXXVI (1981), p. 487.

17 Reiser, 'The Science of Diagnosis', pp. 831–2.

18 Article quoted in George Rosen, 'A Note on the Reception of the Stethoscope in England', *Bulletin of the History of Medicine*, VII (1939), p. 94.

19 V. V., 'To the Stethoscope', *Blackwood's Edinburgh Magazine*, LXI (1847), p. 361.

20 Ibid.

21 Gustavus Hindman Miller, *The Wordsworth Dictionary of Dreams: An Alphabetical Journey through the Images of Sleep* (Ware, Hertfordshire, 1994), p. 531.

22 Thomas Mann, *The Magic Mountain* [1924] (London, 1999), p. 174.

23 Ibid., p. 177.

24 Stanley Joel Reiser, *Medicine and the Reign of Technology* (Cambridge, 1978), p. 37.

25 Ibrahim R. Hanna and Mark E. Silverman, 'A History of Cardiac Auscultation and Some of Its Contributors', *American Journal of Cardiology*, XC/1 (2002), p. 261.

26 Ibid., p. 264.

27 Ibid.

28 Peter R. Fleming, *A Short History of Cardiology* (Amsterdam, 1997), p. 129.

29 John M. Picker, *Victorian Soundscapes* (Oxford and New York, 2003), p. 7.

30 Jonathan Sterne, *The Audible Past: Cultural Origins of Sound Reproduction* (Durham, NC, and London, 2003), p. 23.

31 Picker, *Victorian Soundscapes*, pp. 7–11.

32 Roy Porter, *The Greatest Benefit to Mankind: A Medical History of Humanity from Antiquity to the Present* (London, 1997), p. 354.

33 Ibid., p. 344.

34 Ibid., pp. 344–6.

35 Lachmund, 'Between Scrutiny and Treatment', p. 793.

36 Ibid.

37 Geoffrey Marks, *The Story of the Stethoscope* (Folkstone, 1972), p. 16.

4: Routine

1 Mikhail Bulgakov, *A Country Doctor's Notebook* [1925–7] (London, 1995), p. 47.

2 Ibid., p. 50.

3 Roy Porter, *The Greatest Benefit to Mankind: A Medical History of Humanity from Antiquity to the Present* (London, 1997), p. 678.

4 Ibid., p. 685.

5 John Camm, quoted in Geoff Watts, 'Obituary: Aubrey Gerald Leatham', *The Lancet*, CCCLXXX /9850 (2012), p. 1302.

6 Lydia Wytenbroek, 'In Honor of Nurses Week – The Stethoscope: A Tool of Nurses' Trade since the 1930s', https://batescenterblog.wordpress.com, 12 May 2016.

7 Ibid.

8 David Littmann, 'Stethoscopes and Auscultation', *American Journal of Nursing*, LXXII/7 (1972), p. 1239.

9 Liliana David and Dan L. Dumitrascu, 'The Bicentennial of the Stethoscope: A Reappraisal', *Clujul Medical*, XC/3 (2017), pp. 361–3.

10 Stefan Krebs and Melissa Van Drie, 'The Art of Stethoscope Use: Diagnostic Listening Practices of Medical Physicians and "Auto Doctors"', *ICON: The Journal of the International Committee for the History of Technology*, XX/2 (2014), pp. 92–114.

11 Rudolf Heßler, *Der Selbstfahrer: Ein Handbuch zur Führung und Wartung des Kraftwagens* (Leipzig, 1926), p. 216; as quoted in Krebs and Van Drie, 'The Art of Stethoscope Use', p. 96.

12 Krebs and Van Drie, 'The Art of Stethoscope Use'.

13 René-Théophile-Hyacinthe Laennec, *A Treatise on the Diseases of the Chest and on Mediate Auscultation*, trans. John Forbes, 3rd edn (London, 1827), p. 720.

14 Ibid.

15 Ben De Bruyn, personal communication with the authors, 9 April 2020.

16 William Youatt, *Cattle: Their Breeds, Management and Diseases* (London, 1834), p. 533. Quoted in Ben De Bruyn, *The Novel and the Multispecies Soundscape* (Cham, Switzerland, 2020), p. 186.

17 Ibid.

18 Phil Tomaselli, *Givenchy in the Great War: A Village on the Front Line, 1914–1918* (Barnsley, 2016), p. 140.

19 New Zealand Ministry for Culture and Heritage, 'Listening Underground with a Geophone', https://nzhistory.govt.nz, 22 March 2017. See also J. C. Neill, *The New Zealand Tunnelling Company, 1915–1919* (Auckland, 1922).

20 Dominic Gordon, 'Yarraville Rififi: My Life in Cinema and Crime', *Meanjin*, LXXIX/1 (2020), pp. 30–37.

21 P. J. Bishop, 'Evolution of the Stethoscope', *Journal of the Royal Society of Medicine*, LXXIII (1980), p. 455.

22 Julia Donaldson, *Zog* (London, 2012).

5: Learning

1 Jens Lachmund, 'Making Sense of Sound: Auscultation and Lung Sound Codification in Nineteenth-Century French and German Medicine', *Science, Technology, and Human Values*, XXIV/4 (1999), p. 439.

2 Hubert I. Caplan, 'Stethoscopes as Neckwear', *Archives of Internal Medicine*, CLXIII/21 (2003), pp. 2652–3.

3 Pari Basharat, 'The Stethoscope', *Canadian Medical Association Journal*, CLXXV/7 (2006), p. 780. Reproduced with permission.

4 Anna Harris and Melissa Van Drie, 'Sharing Sound: Teaching, Learning and Researching Sonic Skills', *Sound Studies: An Interdisciplinary Journal*, 5/1 (2015), pp. 98–117.

5 Norm Friesen, *The Textbook and the Lecture: Education in the Age of New Media* (Baltimore, MD, 2018), p. 110.

6 Tim Ingold, *Anthropology and/as Education* (London and New York, 2018).

7 Edward I. Bluth, 'James Hope and the Acceptance of Auscultation', *Journal of the History of Medicine and Allied Sciences*, XXV/2 (1970), p. 204.

8 John Earis and Helen Earis, 'The Tale of an Old Stethoscope: From Mr Bampton (1816–65) to the British Thoracic Society', *Journal of Medical Biography*, XVI (2008), pp. 66–71.

9 T. W. Pocock, 'Letter to the Editor', *London Medical Gazette* (1838), n.s. 2, p. 741. Quoted in Bluth, 'James Hope and the Acceptance of Auscultation', p. 208.

10 P. R. Fleming, *A Short History of Cardiology* (Amsterdam, 1997), p. 99.

11 J. H. Bennett, *An Introduction to Clinical Medicine: Six Lectures on the Method of Examining Patients; Percussion, Auscultation, the Use of the Microscope, and the Diagnosis of Skin Diseases* (Edinburgh, 1853).

12 Ibid.

13 Melissa Van Drie and Anna Harris, 'The Stethoscope Goes Digital: Learning through Attention, Distraction and Distortion', *Gesnerus*, LXXVII/1 (2020), pp. 123–48.

14. Melissa Van Drie, 'Training the Auscultative Ear: Medical Textbooks and Teaching Tapes (1950–2010)', *Senses and Society*, VIII/2 (2013), pp. 165–91.

15 Ibid.

16 Christian Greiffenhagen, 'The Materiality of Mathematics: Presenting Mathematics at the Blackboard', *British Journal of Sociology*, LXV/3 (2014), pp. 502–28.

17 Rachel Vaden Allison, 'The Use of Blackboards and Chalk in Contemporary Anatomy Education', in *Making Sense of Medicine: Material Culture and the Reproduction of Medical Knowledge*, ed. J. Nott and A. Harris (Bristol, 2022).

18 Karin Bijsterveld, *Sonic Skills: Listening for Knowledge in Science, Medicine and Engineering, 1920s–Present* (London, 2019).
19 Van Drie, 'Training the Auscultative Ear', pp. 165–91.
20 Norm Friesen, 'A Brief History of the Lecture: A Multi-Media Analysis', *Medien Padagogik*, XXIV (2014), pp. 136–53.
21 Norm Friesen and Wolff-Michael Roth, 'The Anatomy Lecture Then and Now: A Foucauldian Analysis', *Educational Philosophy and Theory*, XLVI/10 (2014), pp. 1111–29.
22 Michel Foucault, *The Birth of the Clinic* [1963], trans. A. M. Sheridan (London and New York, 2003).
23 Max Peters and Olle Ten Cate, 'Bedside Teaching in Medical Education: A Literature Review', *Perspectives on Medical Education*, III/2 (2014), pp. 76–88.
24 Ibid.
25 Tom Rice, '"Beautiful Murmurs": Stethoscopic Listening and Acoustic Objectification', *Senses and Society*, III/3 (2008), pp. 293–306.
26 Van Drie, 'Training the Auscultative Ear'.
27 Ibid.
28 D. Cugell, *Introduction to Breath Sounds. Audiographic Series 12*, audiocassette (1975), p. 1.
29 S. Lehrer, *Script for Accompanying Tape in Understanding Lung Sounds, with Audiocassette* (Philadelphia, PA, 1984), p. 119.
30 Jeffrey B. Cooper and Viviany R. Taqueti, 'A Brief History of the Development of Mannequin Simulators for Clinical Education and Training', *Postgraduate Medical Journal*, LXXXIV/997 (2008), pp. 563–70.
31 Harry Owen, 'Early Use of Simulation in Medical Education', *Simulation Healthcare*, VII/2 (2012), pp. 102–16.
32 E. William St Clair et al., 'Assessing Housestaff Diagnostic Skills Using a Cardiology Patient Simulator', *Annals of Internal Medicine*, CXVII/9 (1992), pp. 751–6.
33 Lara Lewington, 'The Medical Mannequins "Brought to Life" with Emotions', www.bbc.com, 23 May 2014.
34 Jonathan Sterne, *The Audible Past: Cultural Origins of Sound Reproduction* (Durham, NC, and London, 2003). See also Shigehisa Kuriyama, *The Expressiveness of the Body and the Divergence of Greek and Chinese Medicine* (London, 2002); Bijsterveld, *Sonic Skills*; Projit Bihari Mukharji, 'Akarnan: The Stethoscope and Making of Modern Ayurveda, Bengal, c. 1894–1952', *Technology and Culture*, LX/4 (2019), pp. 953–78.

6: Obsolescence?

1 Andrew Bomback, 'Why Doctors Still Need Stethoscopes', www.theatlantic.com, 10 May 2016.

2 Ibid.

3 Eric Topol, *Deep Medicine: How Artificial Intelligence Can Make Healthcare Human Again* (New York, 2019), p. 302.

4 Bret P. Nelson and Jagat Narula, 'How Relevant Is Point-of-Care Ultrasound in LMIC?', *Global Heart*, VIII/4 (2013), pp. 287–8.

5 'Almost 200 Years Later, Are We Living in the Final Days of the Stethoscope?', www.sciencedaily.com, 23 January 2014; Bahar Gholipour, 'Stethoscopes Could Become Extinct, Doctors Say', www.livescience.com, 23 January 2014; Seung Lee, 'Death of the Stethoscope', http://motherboard.vice.com, 14 July 2015; 'Trusty Stethoscope Faces Threat from Portable Hi-Tech', www.bbc.com, 24 January 2014; Joshua A. Krisch, 'R.I.P., Stethoscope?', www.popularmechanics.com, 24 January 2014.

6 Nelson and Narula, 'How Relevant Is Point-of-Care Ultrasound in LMIC', p. 287.

7 Bernike Pasveer, 'Knowledge of Shadows: The Introduction of X-Ray Images in Medicine', *Sociology of Health and Illness*, XI/4 (1989), pp. 360–81.

8 Ibid., p. 362.

9 Ibid., p. 372.

10 Melissa Van Drie, 'Training the Auscultative Ear: Medical Textbooks and Teaching Tapes (1950–2010)', *Senses and Society*, VIII/2 (2013), p. 172.

11 Pasveer, 'Knowledge of Shadows', p. 374.

12 Barry F. Saunders, *CT Suite: The Work of Diagnosis in the Age of Noninvasive Cutting* (Durham, NC, and London, 2008).

13 Ibid., p. 3.

14 Jos R.T.C. Roelandt, 'The Decline of Our Physical Examination Skills: Is Echocardiography to Blame?', *European Heart Journal – Cardiovascular Imaging*, XXV/3 (2014), p. 251.

15 Jacalyn Duffin, *To See with a Better Eye: A Life of R.T.H. Laennec* (Princeton, NJ, 1998), p. 132.

16 Quoted in Jonathan Sterne, *The Audible Past: Cultural Origins of Sound Reproduction* (Durham, NC, 2003), p. 132. Emphasis in original.

17 Jens Lachmund, 'Making Sense of Sound: Auscultation and Lung Sound Codification in Nineteenth-Century French and German Medicine', *Science, Technology, and Human Values*, XXIV/4 (1999), pp. 419–50.

18 Jonathan Sterne, 'Mediate Auscultation, the Stethoscope, and the "Autopsy of the Living": Medicine's Acoustic Culture', *Journal of Medical Humanities*, XXII/2 (2001), p. 132.

19 Christina M. Agapakis and Sissel Tolaas, 'Smelling in Multiple Dimensions', *Current Opinion in Chemical Biology*, XVI/5–6 (2012), pp. 569–75.

20 Bruno Latour, 'How to Talk about the Body? The Normative Dimension of Science Studies', *Body and Society*, X/2–3 (2004), pp. 205–29.

21 Lorraine Daston and Peter Galison, *Objectivity* (New York, 2010).

22 Abraham Verghese and Sonoo Thadaney, 'Presence: The Art & Science of Human Connection', https://med.stanford.edu, accessed 20 September 2019.

23 Robert J. Adolph, 'In Defense of the Stethoscope', *Chest*, CXIV/5 (1998), pp. 1235–7.

24 Elazer R. Edelman and Brittany N. Weber, 'Tenuous Tether', *New England Journal of Medicine*, CCCLXXIII/23 (2015), p. 2199.

7: Improvisation

1 Dorothy Mull, 'Anthropological Perspectives on Childhood Pneumonia in Pakistan', in *Anthropology in Public Health: Bridging Differences in Culture and Society*, ed. R. A. Hahn (New York, 1999).

2 Claire L. Wendland, 'This Thing, a Stethoscope', in *Making Sense of Medicine: Material Culture and the Reproduction of Medical Knowledge*, ed. J. Nott and A. Harris (Bristol, 2022).

3 Alice Street and Ian Harper, 'Ian Harper's Development and Public Health in the Himalaya', http://somatosphere.net, 4 February 2015.

4 Rebecca Howes-Mischel, '"With This You Can Meet Your Baby": Fetal Personhood and Audible Heartbeats in Oaxacan Public Health', *Medical Anthropology Quarterly*, XXX/2 (2016), pp. 186–202.

5 Gavin Steingo, 'Listening as Life: Sounding Fetal Personhood in South Africa', *Sound Studies*, V/2 (2019), pp. 155–74.

6 Claire L. Wendland, *A Heart for the Work: Journeys through an African Medical School* (Chicago, IL, 2010).

7 Wendland, 'This Thing, a Stethoscope'.

8 Ibid.

9 Felix I. D. Konotey-Ahulu, 'The World Is Round', *British Medical Journal*, CCCXXXVI/1134 (2008).

10 Alexander Pavlosky et al., 'Validation of an Effective, Low Cost, Free/Open Access 3D-Printed Stethoscope', *PLOS ONE*, XII/3 (2018), pp. 1–10. See also Paula Duhatschek, 'How a London Physician Came Up with the Idea for a 3D Printed Stethoscope', www.cbc.ca/news, 15 March 2018.

11 Tom Zimmerman, '7 DIY Farm Hacks', https://modernfarmer.com, 26 August 2014.

12 Graeme Croft, *The Human Body as an Instrument: An Investigation into Its Music*, Masters thesis, Royal Melbourne Institute of Technology, 2015.

13 Axel Volmar, 'Sonic Facts for Sound Arguments: Medicine, Experimental Physiology, and the Auditory Construction of Knowledge in the 19th Century', *Journal of Sonic Studies*, IV/1 (2013).

14 Melissa Van Drie, 'Training the Auscultative Ear: Medical Textbooks and Teaching Tapes (1950–2010)', *Senses and Society*, VIII/2 (2013), pp. 165–91.

15 Richard C. Cabot, 'A Multiple Electrical Stethoscope for Teaching Purposes', *Journal of the American Medical Association*, LXXXI/4 (1923), pp. 298–9; and S. R. Winters, 'Diagnosis by Wireless', *Scientific American*, CXXIV (June 1921), p. 465.

16 A. B. Kinnier-Wilson, L. Fothergill and Selwyn Taylor, 'Some Applications of a New Electronic Stethoscope', *The Lancet* (1956), pp. 1027–8.

17 Lauritz Bredrup Dahl et al., 'Heart Murmurs Recorded by a Sensor Based Electronic Stethoscope and E-Mailed for Remote Assessment', *Archives of Disease in Childhood*, LXXXVII/4 (2002), pp. 297–301.

18 Morton E. Tavel, 'Cardiac Auscultation: A Glorious Past – and It Does Have a Future!', *Circulation*, CXIII/9 (2006), pp. 1255–9.

19 'Technology Revolution in Stethoscope Science Creating Additional Revenues for Manufacturers', *Health Industry Today* (February 2001), p. 5.

20 iStethoscope Pro, 'Description', https://storespy.net, accessed 23 July 2021.

21 Rajiv S. Vasudevan et al., 'Persistent Value of the Stethoscope in the Age of COVID-19', *American Journal of Medicine*, CXXXIII/10 (2020), pp. 1143–50.

22 Danilo Buonsenso, David Pata and Antonio Chiaretti, 'COVID-19 Outbreak: Less Stethoscope, More Ultrasound', *The Lancet*, VII/5 (2020), p. 27.

23 Vasudevan et al., 'Persistent Value of the Stethoscope in the Age of COVID-19', p. 1145.

24 Yuk-Fai Lau, William Wei and Chu-Pak Lau, 'Are Stethoscopes Risky in COVID-19?', *Postgraduate Medical Journal*, XCVI/1137 (2020), p. 431.
25 Lau et al., 'Are Stethoscopes Risky in COVID-19?', p. 431.

Conclusion

1 Shannon Mattern, 'Urban Auscultation; or, Perceiving the Action of the Heart: How We Listen to the City Is as Important as What We Are Listening For', www.placesjournal.org (April 2020), accessed 18 July 2021.
2 Ibid.
3 Ben De Bruyn, *The Novel and the Multispecies Soundscape* (Cham, Switzerland, 2020), p. 605.
4 Ibid., pp. 735–6.

SELECT BIBLIOGRAPHY

Adolph, Robert J., 'In Defense of the Stethoscope', *Chest*, CXIV/5 (1998),
 pp. 1235–7
Bishop, P. J., 'Evolution of the Stethoscope', *Journal of the Royal Society of
 Medicine*, LXXIII/6 (1980), pp. 448–56
Bluth, Edward I., 'James Hope and the Acceptance of Auscultation', *Journal
 of the History of Medicine and Allied Sciences*, XXII/2 (1970), pp. 202–10
Caplan, Hubert I., 'Stethoscopes as Neckwear', *Archives of Internal
 Medicine*, CLXIII/21 (2003), pp. 2652–3
De Bruyn, Ben, *The Novel and the Multispecies Soundscape* (London, 2020)
Duffin, Jacalyn, *To See with a Better Eye: A Life of R.T.H. Laennec*
 (Princeton, NJ, 1998)
Fleming, Peter, *A Short History of Cardiology* (Amsterdam, 1997)
Foucault, Michel, *The Birth of the Clinic: An Archaeology of Medical
 Perception*, trans. Alan M. Sheridan-Smith (New York, 1973)
Howes-Mischel, Rebecca, '"With This You Can Meet Your Baby": Fetal
 Personhood and Audible Heartbeats in Oaxacan Public Health',
 Medical Anthropology Quarterly, XXX/2 (2016), pp. 186–202
King, Lester S., 'Auscultation in England, 1821–1837', *Bulletin
 of the History of Medicine*, XXXIII/5 (1959), pp. 446–53
Lachmund, Jens, 'Between Scrutiny and Treatment: Physical
 Diagnosis and the Restructuring of 19th Century Medical Practice',
 Sociology of Health and Illness, XX/6 (1998), pp. 779–801
—, 'Making Sense of Sound: Auscultation and Lung Sound Codification
 in Nineteenth-Century French and German Medicine', *Science,
 Technology, and Human Values*, XXIV/4 (1999), pp. 419–50
Laennec, René-Théophile-Hyacinthe, *A Treatise on the Diseases of the
 Chest and on Mediate Auscultation* [1829], trans. John Forbes,
 3rd edn (London, 2015)

Nicolson, Malcolm, 'Having the Doctor's Ear in Nineteenth-Century Edinburgh', in *Hearing History: A Reader*, ed. M. M. Smith (Athens, GA, and London, 2004), pp. 151–86

Porter, Roy, *The Greatest Benefit to Mankind: A Medical History of Humanity from Antiquity to the Present* (London, 1997)

Reiser, Stanley Joel, *Medicine and the Reign of Technology* (Cambridge, 1978), pp. 23–44

—, 'The Science of Diagnosis: Diagnostic Technology', in *The Companion Encyclopedia of the History of Medicine*, ed. W. F. Bynum and R. Porter (New York, 1997), pp. 828–51

Rice, Tom, '"Beautiful Murmurs": Stethoscopic Listening and Acoustic Objectification', *Senses and Society*, III/3 (2008), pp. 293–306

—, '"The Hallmark of a Doctor": The Stethoscope and the Making of Medical Identity', *Journal of Material Culture*, XV/3 (2010), pp. 287–301

—, *Hearing the Hospital: Sound, Listening, Knowledge and Experience* (Canon Pyon, 2013)

Steingo, Gavin, 'Listening as Life: Sounding Fetal Personhood in South Africa', *Sound Studies*, V/2 (2019), pp. 155–74

Sterne, Jonathan, 'Mediate Auscultation, the Stethoscope, and the "Autopsy of the Living": Medicine's Acoustic Culture', *Journal of Medical Humanities*, XXII/2 (2001), pp. 115–36

—, *The Audible Past: Cultural Origins of Sound Reproduction* (Durham, NC, and London, 2003)

Van Drie, Melissa, 'Training the Auscultative Ear: Medical Textbooks and Teaching Tapes (1950–2010)', *Senses and Society*, VIII/2 (2013), pp. 165–91

—, and Anna Harris, 'Sharing Sound: Teaching, Learning and Researching Sonic Skills', *Sound Studies: An Interdisciplinary Journal*, I/1 (2015), pp. 98–117

Volmar, Axel, 'Sonic Facts for Sound Arguments: Medicine, Experimental Physiology, and the Auditory Construction of Knowledge in the 19th Century', *Journal of Sonic Studies*, IV/1 (2013)

ACKNOWLEDGEMENTS

We are very grateful for the support of the funding agencies that made our research possible. Anna's research has received funding from the European Research Council under the European Union's Horizon 2020 research and innovation programme (grant agreement no. 678390). Some of her fieldwork received funding from a Dutch NWO Vici Grant entitled 'Sonic Skills: Sound and Listening in the Development of Science, Technology, Medicine (1920–now)' awarded to Karin Bijsterveld (no. 277-45-003). Tom's research was funded through a UK Economic and Social Research Council (ESRC) PhD studentship and also an ESRC Postdoctoral Research Fellowship (project reference ES/F015224/1).

We would like to thank the Sonic Skills team led by Karin Bijsterveld, especially Melissa Van Drie and Stefan Krebs for their collaboration, some of it co-authored with Anna, on stethoscopes.

We would like to thank Michael, Alex and Phoebe at Reaktion Books and anonymous Reaktion reviewers whose comments have helped us to expand our knowledge and enrich the text. Thanks too to Susie Russell, who helped to identify useful texts and images, and Candida Sánchez Burmester for assisting us with references.

Finally, thanks to our fieldwork participants in our studies in Melbourne, London and Maastricht for their time and for sharing their experiences, skills and stethoscopes with us.

PHOTO ACKNOWLEDGEMENTS

The authors and publishers wish to express their thanks to the below sources of illustrative material and/or permission to reproduce it:

Bibliothèque nationale de France, Paris: p. 21; from Richard C. Cabot, *Physical Diagnosis*, 3rd edn (New York, 1905): p. 119 (*right*); David Caird/Newspix: p. 95; Centers for Disease Control and Prevention (CDC): pp. 88 (Debora Cartagena), 91 (Virginia McPheeters); courtesy Graeme Croft: p. 147; Fox Photos/Getty Images: p. 93; Khalil Hamra/AP/Shutterstock: p. 145; drawing by Anna Harris: p. 19; photos Anna Harris: pp. 110, 163; courtesy Barry Issenberg, MD: p. 122; © Kaisu Koski, photos Martijn de Jong: p. 69; photos Yasuhide Kuge, courtesy of Benesse Art Site Naoshima and Fukutake Foundation: p. 117; from R.-T.-H. Laennec, *Traité de l'auscultation médiate et des maladies des poumons et du coeur* (Paris, 1879): p. 28; Library of Congress, Prints and Photographs Division, Washington, DC: pp. 87, 98; Thomas D. McAvoy/The LIFE Picture Collection/Shutterstock: p. 23; from *The Medical Annual: A Year Book of Treatment and Practitioner's Index*, vol. LIII (Bristol and London, 1935): p. 119 (*left*); NBC Universal Photo Bank/Getty Images: p. 160 (*bottom*); John Olson/© 1968, 2021 Stars and Stripes, all rights reserved: p. 99; Osler Library of the History of Medicine, McGill University, Montreal: p. 9; courtesy Royal Veterinary College, London: p. 94; Science Museum, London, photos Wellcome Collection (CC BY 4.0): pp. 6, 54, 58, 63, 79; courtesy Ben Steiner: pp. 157, 158; Wellcome Collection, London (CC BY 4.0): pp. 20, 29, 31, 43, 47, 55, 62, 78, 82, 120, 150; from *William Strang: Catalogue of His Etched Work* (Glasgow, 1906), photo University of California Libraries: p. 80; from Charles J. B. Williams, *The Pathology and Diagnosis of Diseases of the Chest*, 3rd edn (London, 1835): p. 57.

INDEX

Page numbers in *italics* refer to illustrations